Structured Polymer Properties

The Identification, Interpretation, and Application of Crystalline Polymer Structure

ROBERT J. SAMUELS

Hercules Incorporated

A Wiley-Interscience Publication

JOHN WILEY & SONS
New York · London · Sydney · Toronto

Library of Congress Cataloging in Publication Data:

Samuels, Robert J

   Structured polymer properties.

   Includes bibliographical references.
   1. Polymers and polymerization. 2. Crystals.
I. Title.

QD381.S28   547'.84   73-21781
ISBN 0-471-75155-3

Printed in the United States of America

10 9 8 7 6 5 4 3 2 1

This book is dedicated to the memory of my wife
Esta
whose love, understanding, and encouragement
transformed work into play, and discovery
to joy.

Preface

It is the wish of all active minds to be able to develop in a dynamic growth environment—one that is constantly reinvigorated with new and exciting ideas. At present, this type of excitement exists in the field of crystalline polymer physics. With the recent development of new tools, techniques, and ideas, a stimulating, constantly changing environment has been created. Here the fields of chemistry, engineering, and physics are interfaced operationally, leading each day to new insights into the structural dynamics of synthetic and natural polymer behavior.

Historically, all new developments have a temporal evolution such that a large group of people who might participate in the initial period of excitement and growth are excluded from active participation because of poor communication. Thus modern math, which developed in the early 1900s, took more than twenty years to filter down to a larger public; even now its revolutionizing capacity for intellectual derivatization in such fields as social philosophy, science, and engineering is hardly recognized and only superficially applied.

Polymer physics is presently going through an active period of information gathering and idea generation. New techniques for structure identification and a concurrent deeper understanding of the mechanistic relationships between structure dynamics and the observed properties of a polymer have led to a growing excitement over the potential application of these newly emerging views of polymer behavior to practical problems. It is unfortunate but historically expected, however, that this exciting field of both synthetic and natural polymer investigation has been largely isolated to a small coterie of active investigators. Unfortunately, the main body of these investigations is buried in diverse journals, scattered over the years as pieces of ideas and principles. Such fragments have slowly been organized over time in the minds of the investigators, but largely lost to the uninitiated.

This book is an attempt to overcome the historical temporal lag. It is an organized introduction into an area of polymer physics that is undergoing dynamic change today. The book was originally conceived by the author ten years ago as a research program. At that time, much of the structural research in polymer physics was fragmented among specialists in different techniques, and each specialist viewed crystalline polymers from his own frame of reference. It seemed then that a comprehensive examination of a single series of samples by a wide range of techniques was needed to determine areas of information-overlap. At the same time, by relating the structural information obtained from this study to both the fabrication process used to produce the samples and to the end-use properties of the samples after they were fabricated, a comprehensive view of the fabrication-structure-properties chain would be obtained.

The initial study led to new techniques of measurement and revealed new correlations between structure and properties of crystalline polymers. This, in turn, led to the study of different film and fiber processes, and the examination of vastly different polymers such as polypropylene, poly(ethylene-terephthalate), and hydroxypropyl cellulose. Such studies have led to even more general structure-property relations which emphasize an underlying pattern of material behavior.

The research work was organized in a simple uncomplicated manner, creating a format ideally suited for use in an introductory text for graduate students and researchers in polymer physical chemistry, physics, or engineering. Since the same samples are used for demonstrating each technique, it is possible first to introduce the basic elements of the technique and then, by comparing the results with those from other techniques, to illustrate directly the interrelationship between them. Similarly, by extending the work to fabrication process variables and end-use properties the interpretive elements required to correlate this information can be demonstrated.

The field of polymer physics is in its infancy, dynamically moving toward a more stable maturity. For this reason many diverse ideas are actively being pursued, of which this work is one example. It is the author's hope that this introduction to crystalline polymers will allow more students and researchers to enter this actively charged arena and enjoy the intellectual stimulation and satisfaction such an environment offers.

I want to thank Hercules Incorporated and, in particular, Dr. Dwight C. Lincoln for continuing support and encouragement over the years. I also want to thank the following for their assistance in obtaining much of the experimental data presented herein: C. E. Green, M. D. Chris, R. J. Gardecki, J. A. Jackson, B. J. Kocher, A. N. Abbott, and C. W. Hock.

Except for the Preface, Chapter 1, and Section A in Chapter 4, the material for this book has been taken from work of mine published in the *Journal of Polymer Science*. A special acknowledgment is due Marcel-Dekker, Inc., for permission to reproduce my isotactic polypropylene structure-property study as Section A in Chapter 4. I also want to thank John A. Sauer of Rutgers University for permission to use his excellent micrographs for Figs. 1-4, 1-5, 1-11, and 1-12.

Wilmington, Delaware   *Robert J. Samuels*
July 1973

Contents

1 A Look at Polycrystalline Polymer Structure 1

**2 Identification: Techniques for the Characterization of Polycrystalline
    Polymer Structure** 15

A. Introduction 16

B. Molecular Orientation 17

   1. Introduction 17

   2. Wide-Angle X-Ray Diffraction: Determination of the
      Crystalline Orientation Function 22

      a. Introduction 22

      b. A Model for the System 28

      c. Experimental Determination of $\cos^2 \phi_{hkl}$. 32

      d. Application of f_x to Crystallite Orientation
         Processes 37

   3. Sonic Modulus: Determination of the Amorphous
      Orientation Function 41

      a. Introduction 41

      b. A Model for the System 42

      c. Determination of the Sonic Modulus . . 46

      d. Application of the Sonic Modulus to Orien-
         tation Studies 48

   4. Birefringence: Determination of the Total Molecular
      Orientation 51

      a. Introduction 51

      b. Determination of Birefringence . . . 53

      c. Application of Birefringence to Orientation
         Studies 57

   5. Infrared Dichroism: Molecular Structure and Orien-
      tation Parameters 63

      a. Introduction 63

      b. General Infrared Theory 64

      c. Infrared Measurement of the Transition
         Moment Angle 70

      d. X-Ray Diffraction and Determination of the
         Transition Moment Angle 74

C. Spherulite Deformation 82
   1. Introduction 82
   2. Small-Angle Light Scattering (SALS) . . . 89
      a. Undeformed Spherulites 89
      b. Deformed Spherulites 94
         (1) Theory 94
         (2) Photographic SALS Measurements . 99
         (3) Photometric SALS Measurements . 103

D. Summary 109

3 Interpretation: Structural Interpretation of Fabrication Processes . . 114

A. Introduction 114
B. Isotactic Polypropylene Film Fabrication Processes . . 114
   1. Introduction 114
   2. Molecular Anisotropy 115
   3. Interlamellar Anisotropy 121
   4. Spherulite Anisotropy 133
   5. Summary 139

C. Isotactic Polypropylene Fiber Fabrication Processes . . 140
   1. Introduction 140
   2. Process Analysis 141
      a. Spun Fiber 141
      b. Deformation Processes 145
   3. Summary 157

4 Application: Quantitative Correlation of Polymer Structure with
   End-Use Properties 160

A. Isotactic Polypropylene 160
   1. Mechanical Properties 160
      a. Introduction 160
      b. A General Model for the System . . . 161
   2. Failure Mechanics 172
      a. Structural Theory 172
      b. The Fracture Envelope 173
         (1) Rate and Initial Structure Dependence . 173
         (2) Temperature Dependence . . . 176
      c. Failure Distributions 179
         (1) Rate and Initial Structure Dependence . 179
         (2) Temperature Dependence . . . 185

CONTENTS

      d. True Stress Failure Master Curves

          (1) Rate and Initial Structure De

          (2) Temperature Dependence .

      e. Correlation of Fracture in Fibers a

   3. Yield Behavior and Initial Structure .   ·

   4. Tensile Recovery and Initial Structure   ·

   5. Summary   ·   ·   ·   ·   ·   ·

B. Poly(ethylene terephthalate)   ·   ·   ·   ·

   1. Introduction .   ·   ·   ·   ·

   2. Structure–Property Correlations   ·   ·

      a. Tenacity .   ·   ·   ·

      b. Thermal Annealing   ·   ·

          (1) Shrinkage   ·   ·

          (2) Crimp   ·   ·   ·

          (3) Long Spacing   ·   ·

          (4) Small-Angle X-Ray (SAXS)

      c. Dynamic Loss Modulus   ·

      d. Tensile Modulus   ·   ·

   3. Summary   ·   ·   ·   ·   ·

5 Concluding Remarks .   ·   ·   ·   ·   ·

Index   ·   ·   ·   ·   ·   ·   ·   ·

Structured Polymer Properties

1

A Look at
Polycrystalline Polymer Structure

A polycrystalline polymer is a subtle material. Like other types of polymers, it is composed of a single chemical species (the monomer) polymerized into long-chain molecules. It is described in terms of its average molecular weight and molecular-weight distribution, which, as with noncrystalline polymers, will affect its properties. Superficially then, except for the fact that a portion of the polymer is crystalline, these polymers seem to have little to distinguish them as a separate class of note.

On careful scrutiny, however, a polycrystalline polymer will be observed to exhibit quite extraordinary properties. For example, it is possible to bring together four samples of fiber that as a group exhibit the four extremes of mechanical property behavior available to a solid material. Thus, one fiber is very tough and requires a high force to cause it to deform slightly before breaking. A second fiber is ductile. Application of a small force to this fiber causes it to deform, and it continues to deform (pull out) with continued application of a force until it, too, finally breaks. The third fiber is brittle. Application of a small force to this fiber causes it to break immediately. Finally, fiber four is elastic. When pulled it deforms, but when released it returns rapidly to its original configuration.

As fibers go, none of these four types is unique. Many polymers can be found that will have properties similar to those of the four described here. However, our four fibers have all been produced from one polymer. Going one step further, these four fibers, covering the extremes of material property behavior to be expected from a solid, were prepared from a single lot of the polymer, so that the average molecular weight *and* molecular-weight distribution were the same in each of them. Thus a single polymer can be made to manifest the full range of solid-material properties without changing the basic material variables of molecular structure, molecular weight, or molecular-weight distribution.

The inherent quality of a polycrystalline polymer that allows it to act in this unique way can be understood by considering the variety of factors that

1

affect the observed properties of the material. It is often overlooked that, although the fundamental properties of a material are determined by its molecular structure, the bulk properties are controlled by the organizational arrangement of those molecules. This idea is best exemplified on a molecular level by the DNA molecule. Here only four primary base pairs (the components) are required to form all of the genetic code words along the molecule. By simply changing the sequence of base pair combinations (rearranging their organizational structure) all of the genetic characteristics are defined explicitly. It is not the individual property of the base that determines the final property of the gene, but the specific position which that base has in the superstructure of the genetic material. Thus in nature, the components of a system may be looked on as its bricks, with the final characteristics and properties of the house being determined by the manner in which the bricks have been assembled.

In a polycrystalline polymer it is the organizational arrangement *between* molecules rather than the constituent arrangement *along* a molecule that directs its diverse properties. Here the molecules have two options: they may organize either into a rather loose, noncrystalline structure, or into a highly ordered, crystalline one. These molecular structures, the crystalline and the noncrystalline, make up the two primary building blocks whose arrangement determines the observed properties of the polycrystalline polymer.

Each structure, the crystalline and the noncrystalline has its own inherent properties, which differ from those of the other. The average observed property of a material will result as a consequence of how these two structures interact with each other. The nature of the interaction will depend on how the particular phenomena under investigation propagate through the sample, and this in turn will depend on how much of each structure is present and how the structures are arranged in domains. Complicating this organizational picture is the fact that the molecules composing the crystalline and noncrystalline domains are polymeric. For this reason the properties along the molecular axis will be different from those across the molecular axis, and this will impart anisotropic properties to each domain. Therefore, the most important structural parameters for describing a material are the fraction of each structure present and the average anisotropic alignment (orientation) of the molecules in each domain.

This anisotropic two-phase (domain) system undergoes continual rearrangement under the influence of different fabrication processes, tests, and final end-use application. The particular state of the material at any one of these stages will depend on the previous environmental history it has seen. Thus, superficially, one obtains the impression of an infinitely variable structure that is too complex to handle systematically.

This initial impression of unmanageable complexity is misleading, however. Careful examination of the structural variables, their definitions and significance, as well as of the tools available for structural evaluation, shows that through structural studies a quantitative, workable, simple understanding of deformation processes and mechanical properties can be achieved.

Before considering the techniques available for structural studies it is important to have a visual conception of the organizational options available to a crystalline polymer. Isotactic polypropylene is used herein as a model system to describe the common thread that exists between structure, fabrication, and properties. Thus, initially, a qualitative pictorial description of this polymer's organizational character is necessary if a realistic visual working model of the structural elements and their behavior is to be developed.

The term "isotactic" identifies the particular spatial position of the propylene methyl group along the polypropylene carbon-chain backbone (Fig. 1-1). In the isotactic case the methyl groups all fall on the same side of the main chain carbon backbone (Fig. 1-1a). This is in contrast to the syndiotactic case where each alternating methyl group is on the opposite side of the carbon backbone (Fig. 1-1b). Finally, random placement of the methyl groups along the main chain leads to atactic polypropylene (Fig. 1-1c).

The main chain carbon backbone does not lie flat in a plane. Instead, it has a helical conformation in three dimensions (Fig. 1-2). Isotactic polypropylene has a 3_1 helix conformation, which means it takes three monomer units to form the base helix, the fourth monomer unit spatially repeating the 3_1 helix again. As a consequence of this helical conformation, the pendent methyl group, the C–H group, the main chain equatorial $[(C–C)_{eq}]$ bond, and the CH_2 group all make an angle of approximately 72° with the axis of the helix (Fig. 1-2).

When a melt of these helical chains is allowed to cool, some of the molecules will crystallize into folded-chain lamellae while others will not. In this way a two-phase system of crystalline and noncrystalline regions will be formed. Figure 1-3 is a schematic representation of this chain-folded crystal–amorphous-region model. Here the crystalline region is composed of individual crystallites, which are formed from molecules folded back on themselves. The individual single-crystal lamellae are connected to each other through tie molecules so that they will respond cooperatively to a deformation. The amorphous region may consist of the following: molecules whose complex tacticity prevents their crystallization, molecules excluded from the crystal because of their molecular-weight difference, portions of molecules whose complex tacticity prohibits their inclusion in the folded-chain crystals, the disordered fold regions on the surface of the crystal, and tie molecules that meander randomly before participating in another chain-folded crystal.

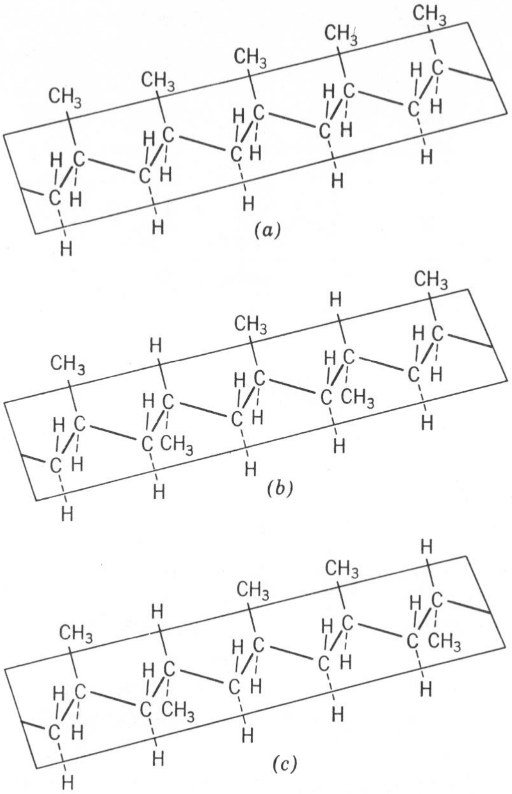

Fig. 1-1 Schematic representation of the spatial disposition of methyl groups in (a) isotactic, (b) syndiotactic, and (c) atactic polypropylene chain segments.

Figure 1-4 is a schematic representation of the molecular arrangement in a single crystal of isotactic polypropylene. A vertical line along the c-axis direction represents the molecular chain axis (helix axis—see Fig. 1-2). If one visually follows a vertical line in Fig. 1-4, it will be seen to bend (a fold), after which the line is then straight again but in the opposite sense. This second straight line will lead to a second fold on the opposite side of the crystal. These connected lines and curves represent the alignment and folding of a single helical polymer molecule. A continuous line looks like a worm with evenly spaced humps and troughs. By folding in this manner, each linear segment of the molecule gets added stability from the van der Waals attraction of the linear segments alongside of it.

The direction of wormlike motion of a molecule in the crystal of isotactic polypropylene is called the a-axis direction. The direction of an individual linear portion is called the c-axis direction (helix or molecular axis), while

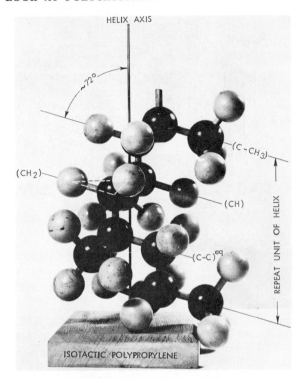

Fig. 1-2 Conformation of isotactic polypropylene. The black balls represent carbon atoms; the gray balls represent hydrogen atoms.

the direction perpendicular to the a-axis, along which the molecules (worms) are arranged beside each other, is called the b-axis direction (Fig. 1-4). The lengths of these three axes and their accompanying angles designate the unit cell or repeating symmetry unit within the crystal lattice. The unit cell within the isotactic polypropylene crystal is outlined by dashed lines in the lower right-hand portion of Fig. 1-4.

From the arrangement of molecules in the crystal it can easily be deduced that the properties of the crystal will vary with crystal direction. Thus if one pulls the crystal in the a-axis direction (the worm-motion direction) and allows it to crack, bundles of aligned molecules are pulled out. Figure 1-5 shows a large single crystal of isotactic polypropylene. The crystal is longest in the a-axis direction because this is its fast-growth direction. The crystal cracked as it settled and bundles of molecules can be seen pulled out into microfibrils along the a-axis direction.

Cracking in the b-axis direction leads to cleavage between molecules (Fig. 1-4) and this should lead to a clean break without any molecules pulled

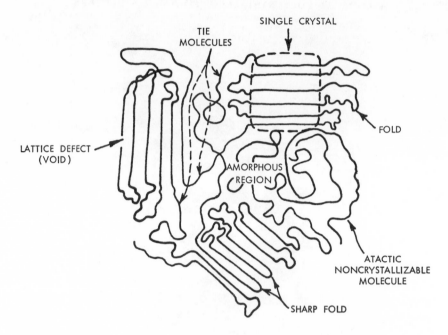

Fig. 1-3 Schematic representation of the single-crystal—amorphous model.

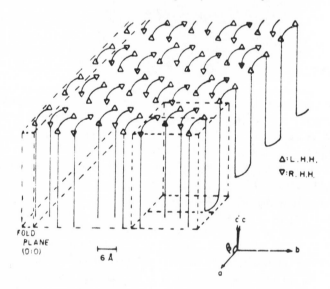

Fig. 1-4 Schematic representation of the molecular arrangement within a single crystal of isotactic polypropylene (J. A. Sauer, Rutgers University, private communication).

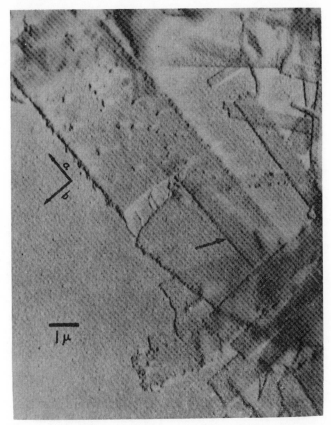

Fig. 1-5 Electron micrograph of a lamellar polypropylene crystal that has fractured parallel to the b axis (J. A. Sauer, Rutgers University, private communication).

out. Such a cleavage is also illustrated in Fig. 1-5, where the cleavage edge is indicated by an arrow.

The crystals and noncrystalline regions are the basic building blocks of isotactic polypropylene. Because of the nature of the crystal nucleation process these regions are organized into supermolecular structures called spherulites. The spherulites are truly large with respect to the components, ranging in size from fractions of a micron to centimeters in diameter.

Figure 1-6 is a photomicrograph of a cross section cut through the center of an isotactic polypropylene spherulite. Rays, like spokes of a wheel, can be seen emanating in all directions from the center of the spherulite. These rays are called radial fibrils. As can be seen in the figure they extend out until they reach a boundary. The boundary is formed when the growing spherulite meets the outward growing front of an adjacent spherulite.

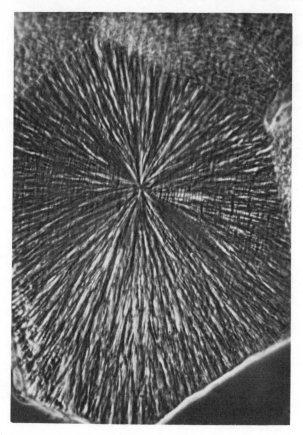

Fig. 1-6 Optical photomicrograph of a cross section of an isotactic polypropylene spherulite.

A crystal nucleation process occurs randomly throughout the crystallizing melt. For this reason the placement of the centers of the spherulites will be located randomly in the bulk sample. A cross section through this material will cut across the center of some of the spherulites and across the tops, bottoms, and intermediate positions of others. For this reason, Fig. 1-6 shows a section cut across the center of the middle spherulite, but not across the center of the adjacent spherulites. Instead, the cut seems to be near their top or bottom so that the ends of the radial fibrils are facing out in these adjacent spherulites.

The crystals and noncrystalline polymer have a definite architecture within the isotactic polypropylene spherulite. The a axis (fast-growth direction) grows radially out from the nucleated center of the spherulite and this results in a specific arrangement of the crystalline and noncrystalline components

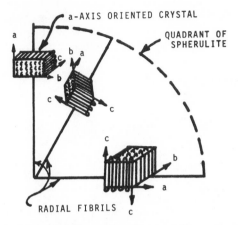

Fig. 1-7   Isotactic polypropylene intraspherulite crystal arrangement.

(see Fig. 1-7). As the spherulite forms, the *a* axis of the crystal is aligned parallel to the radial-fibril axis. Thus, all the crystals along each spoke in the spherulite wheel have a specific alignment with respect to the other crystals along the same spoke, as well as with respect to crystals along the other spokes (fibrils) within the spherulite. The noncrystalline polymer is arranged around the growing crystals and is composed of polymer that cannot enter the crystal lattice. Also, secondary crystallization occurs between the radial fibrils both as the spherulites grow and after they have reached maximum size. This leads to random crystal growth in the region between radial fibrils with no long-range order. Thus the intermediate mortar between the radial fibrils in the spherulite is also composed of crystalline and noncrystalline polymer.

Figure 1-8 schematically summarizes the architectural hierarchy of structures present within an undeformed crystallized sample of isotactic polypropylene. Film is used as the particular example. On the gross scale, the film has a rectangular geometry. Within the film, the sample is seen to be composed of space-filling spherulites. Looking at a particular spherulite, fibrils can be seen arranged radially about its center. Looking at the radial

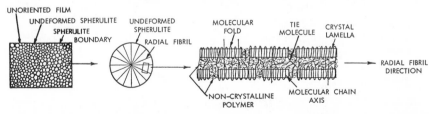

Fig. 1-8   Schematic representation of the structural hierarchy within an undeformed isotactic polypropylene specimen.

fibril in more detail, folded-chain crystals are seen aligned with their *a* axis parallel to the radial-fibril direction along each fibril, with noncrystalline polymer surrounding each crystal. The crystals are seen to be connected both along and between radial fibrils through tie molecules that have entered into the crystallization process of two crystals at the same time. Not shown in the figure are the crystals randomly arranged between the radial fibrils as a consequence of secondary crystallization processes.

When such a complex hierarchy of structure is deformed, rearrangements occur on all structural levels. This is shown schematically in Fig. 1-9, and

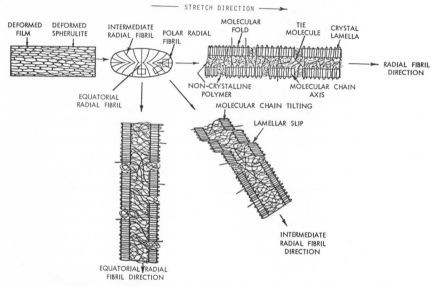

Fig. 1-9 Schematic representation of the structural hierarchy within a deformed isotactic polypropylene specimen.

pictorially in Figs. 1-10 to 1-13. Initially, upon deformation, the spherulites deform affinely as a unit so that the crystals change their orientation as though they were embedded in an elastic sphere. This is followed by a slower process in which there is reorientation of the crystals and noncrystalline regions within the deformed spherulite. The nature of this reorientation will depend on the position of the crystal lamellae with respect to the direction of the deforming stress.

The character of these deformations is shown schematically in Fig. 1-9. When the film is deformed, the spherulites change from a spherical to a spheroidal shape. This process is illustrated with two-dimensional spherulites in Fig. 1-10. The photomicrograph shows a film of isotactic polypropylene that has been deformed. Three large marked spherulites can be seen in the

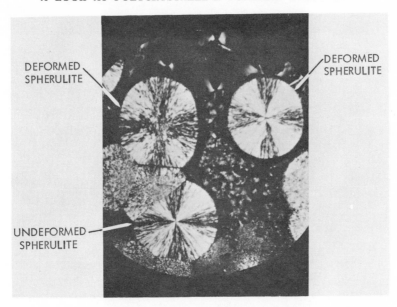

Fig. 1-10 Photomicrograph of deformed isotactic polypropylene spherulites.

picture. The one marked "undeformed spherulite" was under the stretching grips when the bulk film was stretched and did not undergo a deformation; hence it still has its circular shape. The two large spherulites marked "deformed" were stretched with the film. These spherulites originally also had a circular shape but were transformed into an elliptical shape when the film was stretched.

When the spherulite is deformed, the substructure of crystalline and non-crystalline regions reorient, and this reorientation depends on the direction of alignment of the crystals with respect to the deformation direction. As mentioned earlier, when discussing Figs. 1-4, and 1-5, the properties of the crystal will be different in different crystal directions. The manner in which the internal structure changes with deformation is shown schematically in Fig. 1-9. The force is applied horizontally in Fig. 1-9.

In all deformation processes the ultimate orientation of the polymer will result in the helix-chain axis of the molecule becoming aligned parallel to the force direction and, hence, for the crystalline region, all deformation ultimately leads to c-axis alignment along the force direction. In the region of very low extensions ($<0.5\%$) the spherulite deforms elastically as a unit and no permanent internal disruption occurs. At unaxial strains above 0.5%, the radial fibrils aligned parallel to the force direction in the spherulite will have the force concentrated along their axis and focused at the center of the spherulite. The a axis of the crystal is parallel to the force direction along

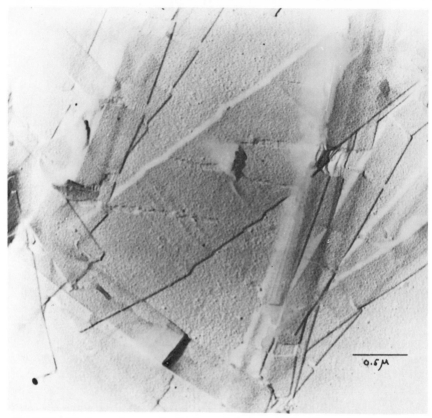

Fig. 1-11 Electron micrograph of deformed single crystals of isotactic polypropylene. The stretch direction is vertical in this figure (J. A. Sauer, Rutgers University, private communication).

these radial fibrils (see Figs. 1-7 to 1-9), and thus this force concentration leads to disruption of the crystals and microfibril formation in the central region of the spherulite. Such a crystal disruption process is shown in Fig. 1-11. Here isotactic polypropylene single crystals lie on a Mylar film that was stretched 50 %. The stretching direction was vertical. Microfibrils can be seen pulled out between sections of crystals whose long axis (a-axis direction) is toward the vertical in this figure.

At angles between 30° and 60° to the applied force, shear forces come into play to facilitate deformation. The shearing forces lead to molecular chain tilting, lamellar slip, and separation of crystallites. This is shown schematically for the intermediate radial fibrils in Fig. 1-9. Pictorially, a large lathlike crystal is seen to lie at a diagonal in Fig. 1-11 (stretch direction vertical). The crystals position is just that required for shear forces to predominate.

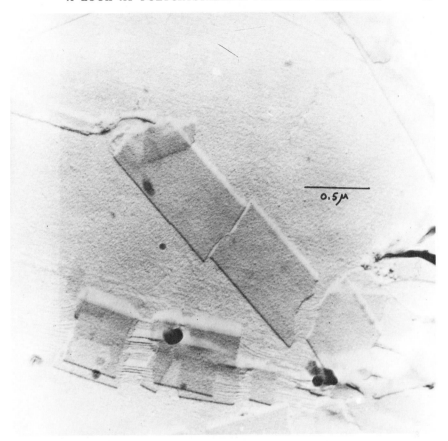

Fig. 1-12 Electron micrograph of deformed single crystals of isotactic polypropylene. The stretch direction is horizontal in this figure (J. A. Sauer, Rutgers University, private communication).

Looking along an edge of this crystal, jogs in the edge alignment can be seen. These jogs are a consequence of the internal molecular-slip processes occurring within the crystal. Such chain-tilting and lamellar-slip processes result in a rotation of the molecular chain axis of the crystallite toward the deformation direction.

The crystallites aligned along the equitorial region of the spherulite have their *a* axis oriented perpendicular to the force direction and their *c* axis in the force direction. Application of a force to these crystals causes them to separate from each other, that is, the distance between their centers increases. Thus, as a force is applied to the sample, the character of the crystal deformation will depend on its orientation with respect to the deforming force.

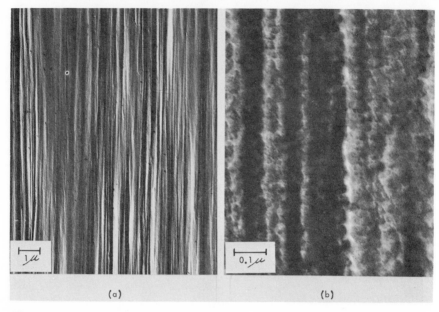

Fig. 1-13 Surface of highly drawn isotactic polypropylene films: (a) untreated (magnification 10,000 ×); (b) acid etched (magnification 150,000 ×).

As the deformation of the spherulites continues, crystal reorganization proceeds by the processes of lamellar slip, orientation, and separation, until the crystal lamellae in all regions of the spherulite become aligned with their c-axis (helix-axis) direction nearly parallel to the deformation direction. As the large crystallites become more highly oriented, crystal orientation becomes exceedingly difficult and further deformation must occur by crystal cleavage. The initial stages of such a cleavage process can be seen within the large diagonal crystal in Fig. 1-11. Figure 1-12 shows isotactic polypropylene crystals on a Mylar film that has been stretched 100% in the horizontal direction. Crystal cleavage has progressed much further in Fig. 1-12 than in Fig. 1-11, for the diagonal crystal in Fig. 1-12 can be seen to have broken into three smaller fragments. Ultimately, as deformation continues, all of the sample becomes highly oriented, crystal cleavage leads to loss of the spherulite structure, and a new microfibrillar structure evolves. Such a microfibrillar structure can be seen in Fig. 1-13. Here a spherulitic film of isotactic polypropylene was drawn to 850% extension. This resulted in a microfibrillated structure. In the untreated sample of this stretched film (Fig. 1-13a) the surface replica shows the film surface is composed of large (~ 2000 Å diameter) fibrils, with no spherulitic character remaining. Upon treatment with nitric acid, which attacks the noncrystalline region surrounding the crystallites, and examining the surface at higher magnification a microfibrillar substructure is resolved (Fig. 1-13b). Here the larger fibrils shown in Fig. 1-13a are seen to be composed of smaller microfibrils, with the crystallites stacked up along the microfibril like corn kernels on a cob.

Thus, as a spherulitic sample is stretched, the spherulites deform, become elongated, and eventually lose their identity, as the crystals yield, orient, and ultimately cleave and stack into microfibrils. Interestingly, the noncrystalline polymer does not seem to be seriously influenced by these complex crystal-deformation processes, but instead the orientation of the noncrystalline region is a linear function of the true strain over the whole draw range (see Chapter 4).

Considering all of the complex processes that occur in a polycrystalline polymer sample during deformation, it is obvious that some method of resolving the structure of each of the domains (crystalline and noncrystalline) individuality is necessary before any sense can be made of the behavior of the sample during processing and testing. This problem of structure identification is considered in Chapter 2, as the different techniques for characterizing polycrystalline polymer structure are described and interrelated.

2

Identification:
Techniques for the Characterization
of Polycrystalline Polymer Structure

A. INTRODUCTION

A knowledge of the morphology of deformed polycrystalline polymers is an essential prerequisite for the development of any theoretical interpretations of deformation mechanisms. The purpose of Chapter 2 is to demonstrate how the morphology of deformed polycrystalline polymers can be characterized in a unified, comprehensive fashion. A two-phase model is developed to describe the polycrystalline state and is subsequently used to characterize the deformation behavior of the polymer. The theoretical foundations of the various independent experimental techniques (e.g., x-ray diffraction, birefringence, sonic modulus, etc.) used to characterize polycrystalline polymer morphology are then developed. Emphasis is placed on the interdependence of the information obtained from these independent experimental techniques as a consequence of the model used to describe the system. The emphasis on interdependence of derived information leads naturally toward new insights into both the molecular and larger morphological characteristics of deformed polycrystalline polymers.

A polycrystalline polymer can be characterized according to phenomena observed at different levels of structure in the sample, that is, the gross sample characteristics, the characteristics of the spherulitic matrix of the sample, or the molecular characteristics of the spherulite substructure. When the sample is deformed, changes occur at all of these morphological levels. As stress is applied, a gross sample deformation occurs. This, in turn, results in spherulite deformation, which consequently leads to a reordering of the crystalline and noncrystalline regions of the polymer. All of these morphological changes must be examined on the same samples if any attempt is to be made to deduce the pattern of the deformation process.

The approach developed here is similar to stop-action photography.

A set of deformed isotactic polypropylene films (designated Series C) is used almost exclusively throughout as the model deformation system. These samples have undergone the simplest deformation process attainable: a continuous, uniform, uniaxial deformation of the polycrystalline film with a corresponding affine deformation of the spherulites. Each of the samples has been deformed to a different extension, however, so that together they correspond to different still sequences of a continuous drawing process. The films, after deformation, were well aged to insure that no changes would occur in the crystal content of the samples over the period in which the large number of measurements were taken. The assumption has been made that, to gain some insight into the complex interactions that occur as a consequence of polycrystalline film deformation, comprehensive examination of the simplest deformation process is potentially most lucrative.

Thus, Chapter 2 is an examination of morphological methods for characterizing deformation mechanisms in polycrystalline polymers arranged in an order that stresses the interdependence of these methods. The examples are designed not only to illustrate the application of each morphological technique but also to demonstrate the diagnostic advantages of examining one material with many different morphological tools.

B. MOLECULAR ORIENTATION

1. Introduction

The present model of the internal structure of a polycrystalline polymer is the chain-folded, crystal–amorphous-region model (Fig. 1-3) in which the individual crystallites are formed from molecules folded back on themselves. The individual single-crystal lamellae are connected to each other through tie molecules so that they will respond cooperatively to a deformation. The noncrystalline region may consist of the following: molecules whose complex tacticity prevents their crystallization, molecules excluded from the crystal because of their molecular-weight difference, portions of molecules whose complex tacticity prohibits their inclusion in the folded crystals, the disordered fold regions on the surface of the crystal, and tie molecules that meander randomly before participating in another chain-folded crystal.

The chain-folded, single-crystal-lamellae–noncrystalline-region model of a polycrystalline polymer is a two-phase system of distinct crystalline and amorphous regions. The amorphous region is characterized in an x-ray diffraction pattern by its diffuse scattering halo as differentiated from the discrete Bragg reflections from the crystalline region. Bonart, Hosemann, Motzkus, and Ruck (1) have pointed out that diffuse scattering need not come only from amorphous polymer but can also arise from imperfections in the crystal lattice. This important observation has led to new insights into the nature of the defect structure of crystalline polymers.

All crystal structures when examined by x-ray diffraction at room temperature have nonideal crystal lattices. Because of this nonideal structure, the observed x-ray diffraction exhibits a decrease in intensity of the crystal Bragg reflections and an increase in the amount of diffuse scatter. As long as the long-range order in the crystal lattice is conserved, only lattice imperfections of the first kind will contribute. The diffuse scatter of the first kind will be caused by deviations in the lattice caused by thermal motion of the atoms in the crystal, frozen displacement or strains, vacancies, dislocations, and other lattice imperfections that cause only short-range disturbances. The general characteristic of distortions of the first kind is that the integral intensity of the Bragg reflections decreases with increasing scattering angle but the reflection widths do not change with scattering angle.

A two-phase system should consist of a well-ordered crystalline region and gradual transitions from the well-ordered crystal to the noncrystalline region should not exist. Regions of intermediate order, "paracrystal," in which long-range order in the crystal is destroyed are characterized as lattice distortions of the second kind. X-ray diffraction from a paracrystalline lattice is characterized not only by a decrease in the integral intensity of the Bragg reflection with increasing scattering angle but by an increase in the reflection width with increasing scattering angle as well. Lattice distortions of the first and second kinds thus manifest different x-ray diffraction behavior. If a polycrystalline polymer is to be considered as a two-phase system of crystalline and amorphous regions, it must first be established that the polymer does not have an appreciable paracrystalline character.

Ruland (2) has demonstrated how x-ray diffraction analysis can differentiate between polymers that can be considered as two-phase systems of crystalline and noncrystalline regions and those that have distinct paracrystalline character. The effect of lattice distortions of all kinds (thermal motion, first kind, and second kind) can be examined in terms of the behavior of a disorder function D expressed as

$$D = \exp\left(-ks^2\right) \tag{2-1}$$

where s is $2\sin\theta/\lambda$, λ is the wavelength of the x radiation, θ is the scattering angle, and k characterizes the disorder and may be represented as

$$k = 0.5B + 0.7a \tag{2-2}$$

where B is a spherically averaged temperature factor taking into account lattice imperfections of the first kind and a is a factor determined by lattice imperfections of the second kind. For a two-phase system of crystalline and noncrystalline regions, only short-range lattice imperfections would be allowed and therefore $k = 0.5B$ for this condition. Thus, Ruland found that isotactic polypropylene must be characterized as a two-phase system (3) since $k = 0.5B$, irrespective of the thermal treatment of the polymer (i.e., the same value of k was found whether the sample was heated to the melting point and then quenched in water at room temperature, or subsequently annealed for 0.5 hr at 160°C, or for 1 hr at 105°C, or if a highly atactic sample was examined). Nylon 6 and 7, on the other hand (2), had varying values of k ($k \neq 0.5B$), depending on the thermal treatment, and these materials had regions of paracrystalline disorder. The boundaries between crystalline and noncrystalline regions in these polymers thus are not well defined. Even this condition does not mean that separate domains do not exist, but that their boundaries are not sharply characterized.

The two-phase model is generally applicable, even for the extreme case of nylon 6 and 7. This is because separate domains exist in these polymers even though their boundaries are not sharply characterized. The model polymer used here is isotactic polypropylene, which has been shown by Ruland to be a two-phase system.

The reason for stressing the two-phase system is that this model is an excellent one for representing the change in properties of a polycrystalline polymer with deformation. Starting with a two-phase model, the observed density, birefringence, sonic modulus, infrared dichroism, and x-ray diffraction data can all be interrelated and cross correlated. The informational agreement between the results of all of these independent physical measurements on a solid polymer is strong support for the validity of the model. Similarly, molecular information obtained with this model agrees with information obtained from other independent methods such as flow birefringence in dilute solution. Finally, no model has been developed which is practically as useful as the two-phase model for interpretation, on the molecular level, of the deformation and mechanical behavior of films and fibers.

How is the two-phase model characterized? What are the most important parameters with which to describe the changes in the observed properties of the polycrystalline polymer when it is deformed? Each phase of the polymer is assumed to have intrinsic properties of its own that are the same as those the polymer phase would have if it existed as a unique, perfectly uniaxially oriented entity. Thus, the crystals will have an intrinsic Young's modulus in the longitudinal and transverse directions, an intrinsic birefringence, characteristic infrared transition moments, and unique x-ray diffraction behavior. The amorphous region will have its own unique intrinsic properties as well. The observed properties of the polycrystalline polymer will be a result of the mixing of these unique phases. In an unoriented polymer, the phases are mixed at random and, hence, any difference in observed properties will be a result of the relative amounts of the two phases present:

$$P_{\text{unoriented}} = \beta P_c + (1 - \beta)P_{\text{am}} \qquad (2\text{-}3)$$

where P is the observed property of the polycrystalline polymer, P_c is the intrinsic property of the crystalline region, P_{am} is the intrinsic property of the amorphous region, and β and $(1 - \beta)$ are fractions of crystalline and amorphous material, respectively.

Since the intrinsic properties of the phases are defined for ideally oriented states, they are inherently anisotropic. For this reason, orienting these phases in a given direction due to some deformation mechanism will lead to a manifestation of the anisotropic character of the phases. Thus, a quantitative measure of the orientation of each phase will be required, whether the observed property of the oriented polymer is a direct measure of the average anisotropy or is a difference measurement of a given property of the oriented polymer relative to the unoriented state. Any observed anisotropic property of the oriented polycrystalline polymer $\Delta P_{\text{oriented}}$ will, therefore, be a function of the properties of each of the phases, and may be expressed as

$$\Delta P_{\text{oriented}} = \beta P_c f_c + (1 - \beta)P_{\text{am}} f_{\text{am}} \qquad (2\text{-}4)$$

where f_c and f_{am} are the orientation functions characterizing the average orientation of the crystalline and amorphous phases, respectively. For example, birefringence, which is an anisotropic property (difference between refractive indexes), can be described by the equation

$$\Delta_T = \beta \Delta_c^0 f_c + (1 - \beta)\Delta_{\text{am}}^0 f_{\text{am}}$$

where Δ_T is the measured birefringence of the oriented polymer and Δ_c^0 and Δ_{am}^0 are the intrinsic birefringence of the crystalline and amorphous regions, respectively. Analogously, the difference between the sonic modulus of an oriented and unoriented sample (ΔE^{-1}) can be described by an equation having the same form:

$$\frac{3}{2}(\Delta E^{-1}) = \frac{\beta f_c}{E_{t,c}^0} + \frac{(1 - \beta)f_{\text{am}}}{E_{t,\text{am}}^0}$$

where $E_{t,c}^0$ and $E_{t,\text{am}}^0$ are the intrinsic sonic lateral moduli of the crystalline and amorphous regions, respectively.

When a specific property is not being considered, but instead a general definition of the average orientation state of the polymer is desired, then eq. 2-4 becomes

$$f_{\text{av}} = \beta f_c + (1 - \beta)f_{\text{am}}$$

Thus, the most important parameters with which to characterize an anisotropic property of an oriented, polycrystalline polymer are the fraction of each phase present, the intrinsic property of each phase, and the orientation function of each phase. In general, the fraction of each phase present in the polymer is determined by a density or x-ray diffraction measurement. For the density determination of the fraction of each phase present, the density of the crystalline and amorphous phases must be known individually. The crystal density d_c can be calculated from the known unit-cell structure and the unit-cell monomer molecular weight, while the amorphous region density d_{am} is obtained from measurements on the pure amorphous polymer when it is available, or by extrapolation of the density–temperature curve of the melted polymer. The fraction of crystals β present in the polymer sample is then given by the relation:

$$\beta = \frac{d_c(d - d_{am})}{d(d_c - d_{am})} \tag{2-5}$$

where d is the measured density of the sample. The density of the polymer sample may be determined by any of a number of standard experimental methods (4) including pycnometry, hydrostatic weighing, and use of a density gradient tube.

The other common method for determining the fraction of crystals present in a polycrystalline polymer sample is x-ray diffraction. This general method utilizes either the intensity relations between crystalline peaks and the amorphous background or the absolute intensity of one of them to determine the amount of crystalline and amorphous material present in the sample (5, 6). The x-ray method of crystallinity determination developed by Ruland (2, 3) accounts for the effects of lattice distortions of the first and second kind through the use of the disorder function, as was discussed above, and this is the most reliable x-ray crystallinity method available at the present time.

The infrared absorption of a polymer can be used to obtain a measure of the fraction of crystals in the sample (7). If separate infrared absorption bands due only to the crystalline and amorphous regions can be identified, the infrared method can be used to obtain absolute values of the fraction of crystals. If, as is more common, absorption bands characteristic exclusively of the crystalline region, but none characteristic exclusively of the amorphous region, are present, the infrared method can be used as a relative measure of the fraction of crystals that then must be calibrated by an absolute method such as density or x-ray diffraction. Since infrared measurements are fast, simple, and sensitive to small changes in crystallinity, they are useful for determining the fraction of crystals even as a relative method.

Attempts have been made to use nuclear magnetic resonance (NMR) techniques to determine the fraction of crystals in a polymer (8). The method has been found to correlate with density and x-ray diffraction measurements for a number of polymers, but only over limited temperature regions. The method is sensitive to the motion of molecules and, hence, restraints imposed by orientation will also influence the observed results. These complicating features of the method, added to the difficulty of obtaining data, make NMR an impractical method for phase-concentration determinations at present.

The major experimental problem in the characterization of the morphology of deformed polycrystalline polymer films has not been the determination of the fraction of each phase present in the two-phase system because some methods for this characterization have been available for many years. Instead, the main problem has been the determination of the intrinsic properties and the orientation function of each phase of the polymer. The general two-phase approach to this problem lay dormant for some time (9), but growing interest in recent years has given new impetus to this area of investigation. Quantitative measurements of the intrinsic properties and orientation functions of the phases of some polyolefin polymers can now be obtained by means of x-ray diffraction, sonic modulus, birefringence, and infrared dichroism. These solid-state measurements not only yield information about the orientation of the molecules in each of the phases, but also provide considerable insight into the structure of the molecules being studied. The purpose of Sections B.2.a–B.5.d is to elucidate how this important morphological information is obtained.

2. Wide-Angle x-Ray Diffraction: Determination of the Crystalline Orientation Function

a. Introduction. In a polycrystalline polymer, the crystalline region is composed of crystallites, whose dimensions can be determined from small-angle x-ray diffraction measurements and from analysis of the breadth of wide-angle diffraction arcs. The crystallites are composed of an ordered array of polymer molecules. The characteristics of the array are described by the unit cell of the crystal. The unit cell is the smallest subunit which, on repetition, will yield the crystal structure. The unit cell quantitatively describes the manner in which the polymer molecules pack into the crystal lattice. Figure 2-1 is a picture of the unit cell of polyethylene, which is orthorhombic with the dimensions shown (10). The orthorhombic unit cell consists of three perpendicular axes of dissimilar lengths.

Let us consider the x-ray diffraction behavior of a single crystal that has an orthorhombic unit cell. If a monochromatic x-ray beam impinges on this crystal, and the crystal is rotated, a diffraction pattern will be obtained that is characteristic of the size and shape of the unit cell. Bragg has shown that if a plane of atoms is regarded as a mirror, that is, reflects a portion of the x-rays at an angle of reflection equal to the angle of incidence, the condition for reinforcement of waves from successive planes of the same kind in the

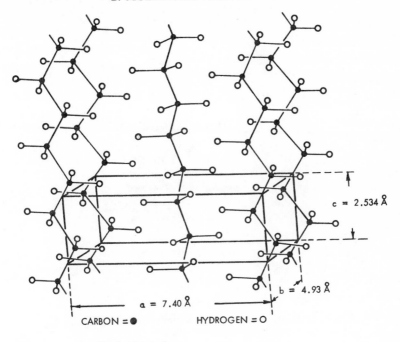

Fig. 2-1 Unit cell of polyethylene.

crystal is that the difference in the lengths of the paths of the reflected rays be equal to an integral number of wavelengths. If θ is the angle between the incident ray and the reflecting plane, the condition for reinforcement of all the reflected rays is

$$n\lambda = 2d \sin \theta \tag{2-6}$$

where $n = 0, 1, 2, 3$, and so on, λ is the wavelength, and d is the interplanar spacing. As the crystal is rotated, the different parallel sets of planes will be brought into a position of reflection and the reinforced reflected beam will impinge on a photographic film and manifest itself as a sequence of black spots (Fig. 2-2). The directions of the reflected beams are governed entirely by the geometry of the lattice (by the orientation and spacings of the planes of atoms). Thus, the size and shape of the unit cell determine the direction the reflected beams will take.

General equations have been derived that characterize the reflections from various unit cells. In the orthorhombic unit cell, the general equation takes the form

$$d_{(hkl)} = \left[\left(\frac{h^2}{a^2} + \frac{k^2}{b^2} + \frac{l^2}{c^2} \right)^{1/2} \right]^{-1} \tag{2-7}$$

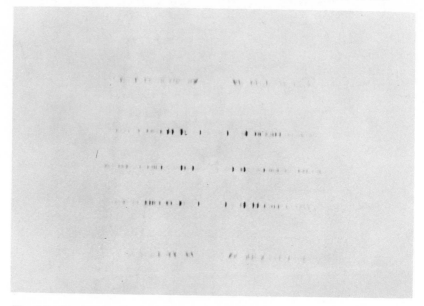

Fig. 2-2   Single-crystal rotation x-ray diffraction pattern of salicylic acid.

or

$$\sin^2 \theta = \frac{\lambda^2}{4}\left(\frac{h^2}{a^2} + \frac{k^2}{b^2} + \frac{l^2}{c^2}\right)$$

(2-8)

Here $d_{(hkl)}$ is the interplanar spacing for the reflecting plane designated by its Miller indices h, k, and l while a, b, and c are the unit-cell dimensions. Other unit-cell structures are characterized by a different arrangement of reflections. Thus, from an analysis of the crystal's diffraction pattern, the unit-cell structure may be determined.

A polycrystalline polymer is not composed of a single crystal but of many small crystallites. Each crystallite will act as a single crystal and contribute reflections to the x-ray diffraction pattern. The resulting wide-angle x-ray diffraction pattern will then be the result of the superposition of the diffraction from each crystallite individually. The resulting pattern is thus a function of the number of crystallites present in the x-ray beam and the orientation (position) of each crystallite relative to the x-ray beam.

If all of the crystallites are randomly oriented in the polymeric film (unoriented sample) and an x-ray beam is sent through the film, some of the crystallites will give no diffraction because none of their planes will be in a reflecting position. Other crystallites will be lying in a position for reflection of a crystal plane; for example, the (110) plane. All of the crystallites that are in the position to reflect the (110) plane will give a reflected beam at the same angle with reference to the incident x-ray beam. The locus of all directions making that angle with the incident beam is a cone having the incident beam as its axis. The resulting cone impinges on the flat x-ray film and appears as a ring on the developed film. This is called the Debye–Scherrer ring and its position on the film characterizes the interplanar spacing that yielded the reflection by means of eq. 2-6. Other crystallites in the polymer film will be in a position to reflect different crystal planes and the sum of all of these reflections will be the final powder pattern obtained from the sample (Fig. 2-3).

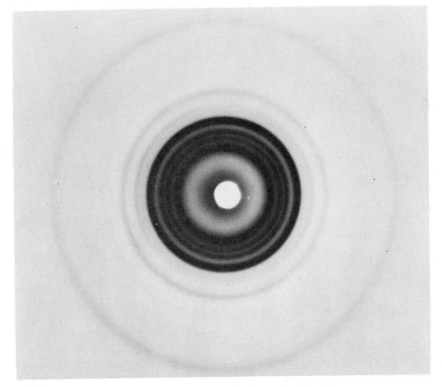

Fig. 2-3   Powder x-ray diffraction pattern of isotactic polypropylene.

If the crystallites are not oriented at random but have their crystal axes oriented around a preferred direction, the Debye–Scherrer ring will not be complete but instead will take the form of an arc. In uniaxial orientation, the crystal axes (such as the c axis) will have a preferred direction with respect to a reference direction in the sample (i.e., the axis of a fiber or the stretch direction of a film). At the same time, the crystal axes will have a random orientation in the plane perpendicular to the reference axis in the sample.

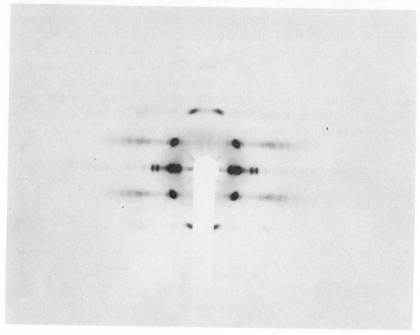

Fig. 2-4   Oriented x-ray diffraction pattern of isotactic polypropylene.

Thus, if one takes a diffraction photograph of a uniaxial sample with the incident x-ray beam parallel to the stretch direction (reference direction) a Debye–Scherrer powder pattern will be obtained (Fig. 2-3). Such a pattern demonstrates that the crystallites are randomly oriented in the plane perpendicular to the reference direction. If one now takes a diffraction photograph of the same uniaxial sample with the incident x-ray beam normal to the stretch direction, an oriented rotation pattern will be obtained (Fig. 2-4). This pattern demonstrates that the crystallites have a preferred orientation with respect to the reference direction. As the crystallites orient more perfectly with respect to the reference direction, the Debye–Scherrer arc will get shorter and shorter. In the limit of perfect orientation, a single-crystal rotation pattern made up of sharp spots (similar to Fig. 2-2) would be obtained. This transition from an unoriented to a highly oriented diffraction pattern is shown more clearly in Fig. 2-5. Here the diffraction pattern is seen to change systematically from an unoriented Debye–Scherrer pattern to a highly oriented pattern with increasing elongation of the original unstretched isotactic polypropylene film.

The foregoing shows that a range of behavior can be expected from the wide-angle x-ray diffraction of a uniaxially oriented polymer film, ranging from a single-crystal rotation pattern to a powder pattern, depending on the

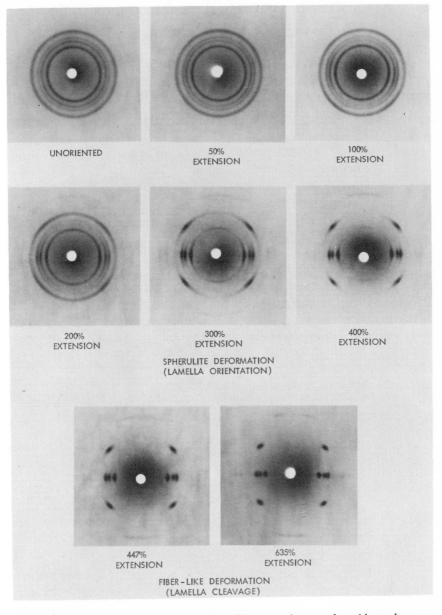

Fig. 2-5 Effect of isotactic polypropylene film extension on the wide-angle x-ray diffraction pattern.

orientation of the crystallites in the polymeric film sample. The average orientation of the crystallites can be formalized on a quantitative numerical basis using Hermans' orientation function f. To do this, one must first be able to determine quantitatively the average angle of orientation a given crystal axis makes with the reference direction. Several models have been proposed as a general solution to this problem and they are reviewed below.

b. A Model for the System.   The orientation function f describes the orientation of the crystallite axis relative to some reference direction in the sample. The orientation function is defined (12) as

$$f_x \equiv \frac{3 \overline{\cos^2 x} - 1}{2} \tag{2-9}$$

where $\overline{\cos^2 x}$ designates the average cosine squared value of the angle x between the reference direction in the sample and the x crystallographic direction.

Stein (13) has set up a generalized model for uniaxial crystal orientation. He represented his coordinate system as shown in Fig. 2-6.

The Z axis of the X, Y, Z Cartesian coordinate system is taken as the stretching direction (reference direction). The angles α, β, and ϵ are measured

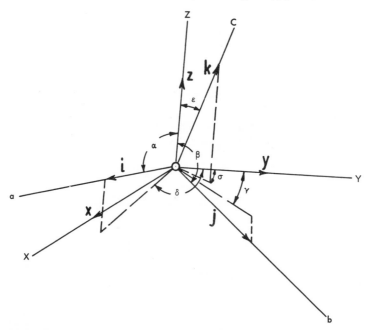

Fig. 2-6   Generalized model for crystal orientation [Stein (13)].

between the Z axis and the a, b, and c crystallographic axes, respectively. Since uniaxial orientation is assumed, and there is cylindrical symmetry about the c axis, the three additional angles δ, γ, and σ between the Y axis (taken to be the plane of the polymer film) and the projection of the three crystallographic axes in the XY plane vary randomly.

The three orientation functions that define the degree of orientation of the three crystallographic axes with respect to the stretching direction are then defined (see eq. 2-9) as

$$f_\alpha = \frac{3\,\overline{\cos^2 \alpha} - 1}{2}$$

$$f_\beta = \frac{3\,\overline{\cos^2 \beta} - 1}{2} \tag{2-10}$$

$$f_\epsilon = \frac{3\,\overline{\cos^2 \epsilon} - 1}{2}$$

If the orientation is random, the crystallographic axes take all directions with equal probability, the value of $\overline{\cos^2 x}$ obtained by averaging over the surface of a sphere with all directions equally probable is $\frac{1}{3}$ and the orientation function is zero. Thus, for completely random orientation, $f_\alpha = f_\beta = f_\epsilon = 0$. If one of the axes is completely oriented with respect to the stretch direction, for example, $x = 0°$, then $\overline{\cos^2 x} = 1$ and $f_x = 1$. If an axis tends to be perpendicular to the stretching direction, for example, $x = 90°$, then $\overline{\cos^2 x} = 0$ and $f_x = -\frac{1}{2}$. The orientation function for each axis can thus vary from a value of $+1$ for parallel orientation with respect to the stretching direction, through a value of 0 for random orientation, to a value of $-\frac{1}{2}$ if the axis is oriented perpendicular to the stretching direction.

Stein developed this model for the study of polyethylene, which has an orthorhombic unit cell, that is, the a, b, and c axes are mutually perpendicular. An orthogonality relationship exists among the three mutually perpendicular directions

$$\overline{\cos^2 \alpha} + \overline{\cos^2 \beta} + \overline{\cos^2 \epsilon} = 1 \tag{2-11}$$

and therefore

$$f_\alpha + f_\beta + f_\epsilon = 0 \tag{2-12}$$

Thus, for this model, the three orientation functions are not independent and only two of them are required to characterize the orientation.

The orthogonality relationship that makes this model so flexible demands that the a, b, and c crystal axes must be perpendicular and, hence, the unit cell of the material being investigated must be of the isometric, tetragonal, or orthorhombic system.

Wilchinsky (14) showed how this model could be extended to a more general treatment in which $\overline{\cos^2 x}$ can be determined, even if there are no reflecting planes normal to the chosen crystallographic direction. Figure 2-7 illustrates the model in its more general form. Here, the vector $\mathbf{Z}$ represents the stretch direction (reference direction, see Fig. 2-6). OC, OU, and OV represent a Cartesian coordinate system. The desired crystallographic direction is then fixed to coincide with the OC axis. Within this restriction, the crystal may be fixed at any arbitrary position with respect to the OU and OV axes. The position of the crystal, with respect to the reference direction $\mathbf{Z}$ is then specified by the angles α, β, and ϵ as indicated and

$$\mathbf{Z} = (\cos \alpha)\mathbf{i} + (\cos \beta)\mathbf{j} + (\cos \epsilon)\mathbf{k} \qquad (2\text{-}13a)$$

$\mathbf{N}$ is a unit vector along the normal to a set of reflecting planes in the crystal. An (hkl) reflecting plane is represented by triangle abc in Fig. 2-7, where a, b, and c are the crystal axes directions. The vector $\mathbf{N}$ can be described in terms of its components along the OC, OU, and OV axes, that is,

$$\mathbf{N} = (\cos E)\mathbf{i} + (\cos F)\mathbf{j} + (\cos G)\mathbf{k}$$
$$\mathbf{N} = e\mathbf{i} + f\mathbf{j} + g\mathbf{k} \qquad (2\text{-}13b)$$

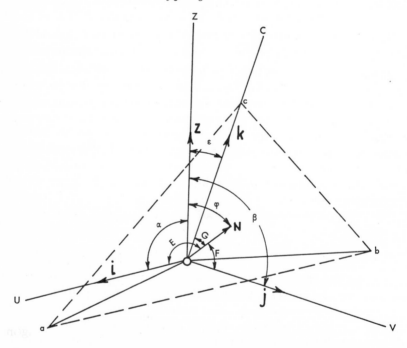

Fig. 2-7   Generalized model for crystal orientation [Wilchinsky (14)].

where $e = (\cos E)$, $f = (\cos F)$, and $g = (\cos G)$ are the direction cosines of N with respect to the OC, OU, and OV axes and $\mathbf{i}$, $\mathbf{j}$, and $\mathbf{k}$ are unit vectors along these axes.

The angles α, β, and ϵ now represent the angles the OC, OU, and OV axes make with the reference direction. Angle ϵ also represents the angle between the chosen crystallographic direction and the reference direction in the sample. The same orthogonality relationship defined in Stein's model (eq. 2-11) holds for the newly defined Cartesian coordinate system OC, OU, and OV, for example,

$$\overline{\cos^2 \epsilon} + \overline{\cos^2 \alpha} + \overline{\cos^2 \beta} = 1$$

The angle between the plane normal N and the reference direction Z is ϕ. From vector algebra $\cos \phi = N \cdot Z$ and thus

$$\cos \phi = N \cdot Z = e \cos \alpha + f \cos \beta + g \cos \epsilon$$

The desired average is then given by the expression

$$\overline{\cos^2} \phi_{hkl} = e^2 \overline{\cos^2 \alpha} + f^2 \overline{\cos^2 \beta} + g^2 \overline{\cos^2 \epsilon} + 2ef \overline{(\cos \alpha)(\cos \beta)}$$
$$+ 2fg \overline{(\cos \beta)(\cos \epsilon)} + 2ge \overline{(\cos \epsilon)(\cos \alpha)} \quad (2\text{-}14)$$

The quantity $\overline{\cos^2 \phi}$ for the planes (hkl) is determined experimentally from the distribution of diffracted intensity from these planes, while the coefficients e, f, and g are calculated from a knowledge of the crystal structure.

Since there are six unknowns in eq. 2-14, the desired quantity $\overline{\cos^2 \epsilon}$ must be evaluated by solving six simultaneous equations involving the six unknowns. Thus, eq. 2-14 evaluated for five sets of reflecting planes having normals in different directions from each other would contribute five of the required equations while the orthogonality relation, eq. 2-11, would provide the sixth.

Sack (15), using a model equivalent to Wilchinsky's, gave a more detailed and complete theory for the problem of calculating f_x for a prescribed direction in the lattice from the known values of $\overline{\cos^2 \phi}$ in other directions. His solution is equivalent to Wilchinsky's. He does, however, derive some simpler formulas, which are valid if certain symmetry conditions are satisfied. He also considers modifications to the solution that would arise from such experimental limitations as overlap of several Debye–Scherrer rings, limited accuracy in measurements, or, if more data are available than would ideally suffice to calculate the required solution.

From an analysis of symmetry conditions, Sack determined that, for a macroscopically uniaxial aggregate of orthorhombic crystals in which the OC, OU, and OV axes are identified with the crystal axes, the minimum number of independent values of $\overline{\cos^2 \phi}$ that must be determined is two. This system is equivalent to Stein's model (Fig. 2-6) and shows it to be a special case of the more general treatment. For tetragonal and hexagonal lattices, Sack finds that the minimum number of independent values of $\overline{\cos^2 \phi}$ is one.

This new formalism makes it possible for the investigator to consider studies of the effects of deformation and processing on the properties of polymers, which he could not do quantitatively before. An excellent example is polypropylene. Natta (16) has shown that isotactic polypropylene has a monoclinic unit-cell structure. Since the b axis of polypropylene is perpendicular to its (040) planes, its orientation relative to the reference (stretch) direction could be measured directly, that is, $\overline{\cos^2 \beta} = \overline{\cos^2 \phi_{040}}$, where β is the angle between the stretch direction and the b axis of the crystal.

The c axis of the isotactic polypropylene unit cell is parallel to the helical axis of the isotactic polypropylene molecule in that cell. Determination of the average orientation of the c axis with respect to the reference direction is thus a determination of the average orientation of the helical molecules of the crystalline region with respect to the reference direction. Since the three crystallographic axes in a monoclinic unit cell are not mutually perpendicular (the angle between the a and c axis in isotactic polypropylene is 99°20′) (16) and no (00l) pure c-axis reflections are present in the x-ray diffraction pattern, Wilchinsky's generalized approach to the determination of the c-axis orientation given above must be used.

Wilchinsky (17) derived from symmetry considerations the following expression relating the variation in intensity with azimuthal angle of the (110), (040), and (130) reflections to the orientation function parameter which characterizes the c axis,

$$\overline{\cos^2 \epsilon} = 1 - \frac{(1 - 2 \sin^2 \rho_2)(\overline{\cos^2 \phi_1}) - (1 - 2 \sin^2 \rho_1)(\overline{\cos^2 \phi_2})}{\sin^2 \rho_1 - \sin^2 \rho_2} \qquad (2\text{-}15)$$

where the subscripts refer to the respective planes, ρ is the angle between the plane normal and the b axis and is calculated from unit cell dimensions, and ϵ is the angle between the reference direction and the c axis. Thus, the orientation function for the c axis of isotactic polypropylene could be calculated once $\overline{\cos^2 \phi_{hkl}}$ had been determined for two reflecting crystal planes [e.g., the (040) and (1$\bar{1}$0) planes].

c. Experimental Determination of $\overline{\cos^2 \phi_{hkl}}$. Once the unit cell of a polymer crystal is known and the reflections have been indexed (i.e., the hkl Miller index assignments are known for each reflection) $\overline{\cos^2 \phi_{hkl}}$ can be determined experimentally. The discussion that follows will be general but the example used will be data from uniaxially oriented, isotactic polypropylene film. Since a uniaxial system is being used, the azimuthal intensity distribution need only be measured with normal incidence of the x-ray beam or θ_{hkl} incidence of the x-ray beam. If the sample did not have cylindrical symmetry in the xy plane (i.e., were not randomly oriented in the plane perpendicular to the stretching direction), the azimuthal intensity distribution of a given plane would have to be measured at various tilt angles of the sample (i.e., a pole-figure analysis of each reflecting plane would be required). In this discussion, we will treat only the uniaxial orientation case. The biaxial case is reviewed by the author elsewhere (18).

Two methods are available for determining the intensity distribution around a Debye–Scherrer arc (i.e., an azimuthal scan). One may photograph the x-ray diffraction pattern of the sample and then examine the desired reflection with a microphotometer, or one may use a diffractometer and measure the intensity distribution directly. The diffractometer technique has the advantage of greater accuracy and flexibility and it is this technique that is used in the experimental examples.

In practice, the diffractometer is set at the radial $(2\theta_{hkl})$ angle of the desired hkl plane to be examined. The sample is then rotated slowly through an angular range large enough to include the azimuthal angles $\rho = 0$–$90°$. Here ρ, the azimuthal angle, is defined as the angle between the stretching direction and the plane of measurement of θ_{hkl}. The intensity at each angle ρ is recorded directly in volts on a strip chart. These intensity values can be converted directly to counts/sec by a suitable scale factor correction. Figure 2-8 is a typical azimuthal trace of the b-axis reflection [(040) plane] from oriented isotactic polypropylene film. The measured intensity I_{meas} in counts per second at each azimuthal angle ρ must be corrected for absorption, polarization, background, and incoherent scattering before it can be used to evaluate $\overline{\cos^2 \phi_{hkl}}$. The equation used for correction of the measured intensity is

$$I(\rho) = (I_{\mathrm{meas}} - I_{\mathrm{background}})K_{\mathrm{polarization}}K_{\mathrm{absorption}} - I_{\mathrm{incoherent}} \qquad (2\text{-}16)$$

The background intensity $I_{\mathrm{background}}$ arises from electronic circuit noise, air scattering, cosmic radiation, and so on, and is determined experimentally by measuring the scattered intensity with no sample in the beam over the complete azimuthal angular range at the radial $(2\theta_{hkl})$ angle being investigated. The unpolarized incident x-ray beam is partially polarized by the sample and this polarization effect is corrected for by the expression

$$K_{\mathrm{pol}} = \frac{2}{1 + \cos^2 (2\theta_{hkl})} \qquad (2\text{-}17)$$

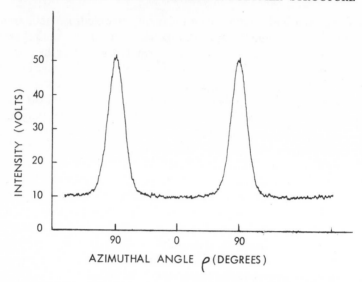

Fig. 2-8 Typical 040 azimuthal intensity distribution scan from oriented isotactic polypropylene.

The absorption correction depends on the angle of incidence of the x-ray beam relative to the polymer film surface. Gingrich (19) has shown that for the incident beam normal to the film surface

$$K_{abs} = \frac{\mu t(\sec 2\theta_{hkl} - 1)}{1 - \exp\left[-\mu t(\sec 2\theta_{hkl} - 1)\right]} \tag{2-18}$$

where μ is the linear absorption coefficient and t is the thickness of the sample film. If the incident beam is inclined at the angle θ_{hkl} to the polymer film surface, the absorption correction has the form

$$K_{abs} = \frac{\exp\left(\mu t \sec \theta_{hkl}\right)}{\sec \theta_{hkl}} \tag{2-19}$$

The final correction is for incoherent scattering (Compton scattering), which originates from the collision of the x-ray beam with loosely bound or free electrons. This correction is generally small in the low-radial-angle region of the radiation intensity curve. The incoherent scattering intensity is described by the expression

$$I_{inc} = C_{inc}\left[\sum_i \left(Z_i - \sum_j f_j^2\right)\right] \tag{2-20}$$

where Z_i is the atomic number of the atom of type i, f_j^2 is the atomic form factor for this atom at the angles of measurement, where the sum over j is over all of the electrons of the atom, and C_{inc} is an experimentally determined, angularly independent constant. This constant is evaluated by assuming that at a sufficiently large radial angle (i.e., 50°) all of the scattering is incoherent, then at that radial angle (20)

$$I_{inc}(50°) = (I_{meas} - I_{background})K_{pol}K_{abs} \tag{2-21}$$

C_{inc} is then evaluated by equating eqs. 2-20 and 2-21.

Substituting the above corrections into eq. 2-16, one obtains the following final equation for the corrected intensity $I(\rho)$ in the case of normal incidence of the x-ray beam:

$$I(\rho) = \left\{ (I_{meas} - I_{background}) \left(\frac{2}{1 + \cos^2 2\theta_{hkl}} \right) \right.$$
$$\left. \times \left[\frac{\mu t(\sec 2\theta_{hkl} - 1)}{1 - \exp\left[-\mu t(\sec 2\theta_{hkl} - 1)\right]} \right] \right\} - I_{inc} \tag{2-22}$$

Since the radial angle is held constant (at $2\theta_{hkl}$) during an azimuthal scan, the polarization, absorption, and incoherent scattering corrections are constant over the whole azimuthal angular range.

The measured intensities need not be obtained in volt readings from a strip chart, but in more sophisticated, automatic x-ray diffractometers may be obtained directly in counts per second at fixed angular intervals by use of a step scanner. The data are then recorded directly on punched tape which is fed to a computer for analysis. This system of data collecting and analysis has the advantage of increased accuracy, flexibility, safety, and efficiency. Figure 2-9 is a photograph of an automatic x-ray diffraction apparatus in use in the author's laboratory (21). A is the automatic controller and logic system. Angular intervals, angular range, and types of motions are programmed on this panel. B is the diffractometer intensity accumulator system. The system can accommodate intensity measurements in the form of counts for a fixed time or time for a fixed count as well as continuous scan. C is the automated goniometer with step-scanning facilities. D is a digital tape punch, which records the intensity data from B in machine language, while E is a digital printer, which will type out the intensity data from B simultaneously with the operation of the tape punch D, in lieu of the tape punch, or remain off during the data-gathering operation. This diffractometer can obtain the pole figure of three independent reflections (i.e., at three $2\theta_{hkl}$ positions) automatically, during one complete operation. Thus, the azimuthal scans of as many as three reflecting planes [e.g., the (040), (110), and (130) crystal planes of isotactic polypropylene] for a uniaxially oriented polymer film can be obtained automatically in this instrument.

Fig. 2-9 Automated x-ray diffractometer for orientation studies.

For a system that is oriented uniaxially

$$\overline{\cos^2 \rho_{hkl}} = \frac{\int_0^{\pi/2} I(\rho) \cos^2 \rho \sin \rho \, d\rho}{\int_0^{\pi/2} I(\rho) \sin \rho \, d\rho} \tag{2-23}$$

Equation 2-23 can be evaluated graphically by multiplying the corrected intensities by the appropriate values of the sine and cosine functions and plotting the results against the azimuthal angle. The areas under the curves can then be measured with a planimeter and used to determine $\overline{\cos^2 \rho_{hkl}}$. As mentioned above, a much more rapid and reliable method of calculating $\overline{\cos^2 \rho}$ is to allow a computer to numerically calculate the value from the input raw intensity data.

Polyani (22) has shown from geometrical considerations that for normal incidence of the x-ray beam $\overline{\cos \phi_{hkl}} = (\overline{\cos \theta_{hkl}})(\overline{\cos \rho_{hkl}})$ and hence

$$\overline{\cos^2 \phi_{hkl}} = (\overline{\cos^2 \theta_{hkl}})(\overline{\cos^2 \rho_{hkl}}) \tag{2-24}$$

If θ_{hkl} incidence of the x-ray beam is used, however, then

$$\overline{\cos^2 \phi_{hkl}} = \overline{\cos^2 \rho_{hkl}} \qquad (2\text{-}25)$$

Once $\overline{\cos^2 \phi_{hkl}}$ has been determined experimentally, the orientation function of the desired crystal axis can be calculated. If a pure axial crystal plane reflection is available, $\overline{\cos^2 X}$ for that reflection is equal to $\overline{\cos^2 \phi_x}$, that is,

$$\overline{\cos^2 \phi_{h00}} = \overline{\cos^2 \alpha}$$

$$\overline{\cos^2 \phi_{0k0}} = \overline{\cos^2 \beta} \qquad (2\text{-}26)$$

$$\overline{\cos^2 \phi_{001}} = \overline{\cos^2 \epsilon}$$

These values can be used directly in eq. 2-10 to calculate the orientation function. If a pure axial crystal-plane reflection is not available in the x-ray diffraction pattern, Wilchinsky's method of analysis (eq. 2-14) must be used. This was illustrated earlier for isotactic polypropylene film. Thus, from an analysis of the azimuthal intensity distribution of the (040) plane $\overline{\cos^2 \phi_{040}}$ is obtained. Since this is an (0k0) pure b-axis reflection, $\overline{\cos^2 \phi_{040}} = \overline{\cos^2 \beta}$. However, since no (001) pure c-axis reflections are present in the x-ray diffraction pattern of isotactic polypropylene, $\overline{\cos^2 \epsilon}$ could not be obtained in this simple manner. Instead, Wilchinsky had to resort to eq. 2-14 which reduced to eq. 2-15 when the symmetry characteristics of the monoclinic unit cell were considered. The angle ρ in eq. 2-15 was calculated for the isotactic polypropylene crystal from a geometric analysis of the unit-cell structure. The values of ρ for the (040), (110), and (130) diffraction planes were calculated to be (17)

$$\rho_{(040)} = 0°$$

$$\rho_{(130)} = 46.7°$$

$$\rho_{(110)} = 72.5°$$

Then $\overline{\cos^2 \epsilon}$ could be calculated by substituting the appropriate ρ_{hk0} with the experimental $\overline{\cos^2 \phi_{hk0}}$ for two reflections into eq. 2-15. The f_ϵ value for the isotactic polypropylene sample is then determined by substituting $\overline{\cos^2 \epsilon}$ into eq. 2-10 and solving.

d. Application of f_x to Crystallite Orientation Processes. Deformation of a polycrystalline polymer film results in an orientation of both crystalline and amorphous regions with respect to the deformation direction. The change in the orientation of the crystalline region can be followed quantitatively by determining the change in the orientation functions for the crystal axes as a function of the deformation of the sample. Similarly, if oriented polymer film is heat treated at fixed elongation, or shrunk, or solvent swollen, or undergoes any other physical treatment, resulting changes in the orientation of the crystallite can be followed quantitatively by studying the changes in the crystal orientation function. Wide-angle x-ray diffraction has been used to determine the crystal orientation function of cellulose (9), polyethylene (23, 24), isotactic polypropylene (18, 23, 25–28), poly-1-butene (23), polyethylene terephthalate (29–32), and hydroxypropylcellulose (33).

The changes that occur in the orientation of the crystallites as a sample is deformed uniaxially can be represented graphically by an orientation function triangle diagram (13). The character of this triangle plot results from the model used to characterize $\overline{\cos^2 \phi_{hkl}}$. The orthogonality of the axes of the model results in the interdependence of the three orientation functions (see eq. 2-12). As a result of the model, only two orientation functions are required to characterize the average orientation of all three axes. Therefore, the state of orientation may be represented as a point on a two-dimensional plot with two of the f's as coordinates.

Figure 2-10 is an orientation-function triangle diagram representing data

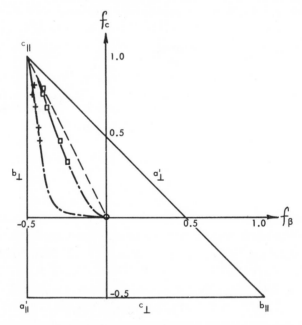

Fig. 2-10 Orientation function triangle diagram for the b and c axes of isotactic polypropylene (25): (☐) Hot-drawn Series C films; (+) melt-spun fibers; (---) cold-drawn films (23).

from samples of isotactic polypropylene that have undergone different physical treatments. Because of restrictions on the range of values imposed by eq. 2-12, all points must occur within the indicated triangular area. When a sample film is unoriented, all of the crystallites are randomly arranged in the film and, hence, $f_\alpha = f_\beta = f_\epsilon = 0$ (in Fig. 2-10, $f_c = f_\epsilon$). Thus, the origin in Fig. 2-10 represents the unoriented state of the material. When one of the crystal axes is oriented parallel to the stretch direction, $f_x = 1.0$ and, therefore, each apex of the triangle diagram represents a state of perfect orientation of one of the three axes along the stretch direction. For example, the apex $c_\parallel$ in Fig. 2-10 corresponds to a state where $f_\epsilon = 1, f_\beta = f_\alpha = -\frac{1}{2}$. This is a state of orientation where the c axis of the crystal lies in the stretching direction and the b and a' axes* are oriented perpendicular to the stretching direction. One further characteristic of the orientation function triangle is that the center of each side of the triangle diagram corresponds to a state of orientation where the crystal axis lies perpendicular to the stretching direction. Thus, the point where the b axis of the crystal is perpendicular to the stretching direction and the other two axes are random is in the center of the side marked $b_\perp$.

The importance of the orientation-function triangle diagram is that one can describe uniaxial orientation processes in terms of the path of a point, which corresponds to a given state of average orientation of the crystallites in the sample around this triangle. To illustrate the power of this form of representation of uniaxial orientation processes, let us examine the orientation behavior of the isotactic polypropylene samples. For isotactic polypropylene, the c axes of the crystal corresponds to the helical chain axis of the molecule. Generally, when a sample is deformed, the polymer chain axis orients in the direction of the deformation. Thus the c axis of the crystal orients in the direction of the deformation in cold-drawn, isotactic polypropylene. The resulting x-ray diffraction pattern for c-axis orientation is shown in Fig. 2-11a. Both melt-spun, isotactic polypropylene fibers (34) and cold-drawn fibers annealed close to the melting point (35) exhibit two types of crystal orientation simultaneously. Most of the crystals exhibit the expected c-axis orientation but some of the crystals exhibit a'-axis orientation. In a'-axis orientation, the polymer-chain axis is nearly perpendicular to the stretching direction and the a' axis of the crystal is oriented toward the stretch direction. This type of orientation appears in the x-ray diffraction

---

* Because of the monoclinic unit-cell structure of isotactic polypropylene, the a axis of the crystal does not fall on the OV axis of Wilchinsky's Cartesian coordinate system (Fig. 2-7). For this reason, f_α does not represent the true a-axis orientation of the crystal, but instead represents the orientation of the OV axis. This axis is designated as the a' axis and is related to the true a axis through the deviation of the β monoclinic angle from 90°. The β angle equals 99°20′ (16) for isotactic polypropylene and, thus, the a'-axis orientation is close to the true a-axis orientation.

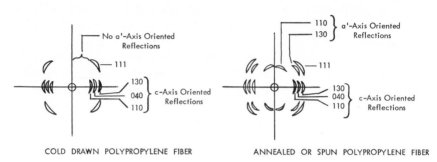

Fig. 2-11   X-ray diffraction diagrams of isotactic polypropylene fibers.

pattern as shown in Fig. 2-11*b*. Thus, *a'*-axis orientation seems to be the high-temperature form in oriented isotactic polypropylene.

The curves in the orientation function triangle diagram (Fig. 2-10) represent data from (1) cold-drawn, isotactic polypropylene film (23), (2) hot-drawn (110°C), Series C isotactic polypropylene film (25), and (3) melt-spun, isotactic polypropylene fibers (35). In this diagram, $f_c = f_\epsilon$, the change in the designation indicating that this is the orientation function representing the orientation of the molecular helical chain axis in the crystalline region of the polymer. The orientation function triangle diagram of the data obtained from cold-drawn, isotactic polypropylene film is a straight line from the origin to the apex $c_{\parallel}$ of the triangle. Movement of a point from the origin to the apex of the triangle is movement in the direction of increasing elongation of the sample. The straight line from the origin to the $c_{\parallel}$ apex for the cold-drawn, isotactic polypropylene fibers indicates that the a' and b axes of the crystallite rotate randomly about the c axis as the c axis is turned toward the stretching direction.

The curve for the melt-spun fibers in Fig. 2-10 (+ points) is bowed toward the $b_{\perp}$ side of the orientation function triangle diagram. The f_c values for these samples were obtained from azimuthal intensity distribution scans of the (040) reflection and the bimodal (110) reflection, and application of the resulting data to eqs. 2-15 and 2-10. The results plotted in the orientation function triangle diagram indicate that the b axis of the crystal orients toward the perpendicular to the spinning direction (fiber axis) faster than the a' axis does, as the c axis orients in the fiber axis (deformation) direction. Examination of the x-ray diffraction pattern (Fig. 2-11*b*) indicates that this a'- and b-axis behavior is observed because a fraction of the crystallites have the a' axis oriented toward the deformation direction, while the rest have the

a' axis oriented perpendicular to the stretch direction. Thus, the average a'-axis orientation of all the crystallites is effectively orienting more slowly toward the perpendicular to the deformation direction than is the b axis of the crystallites.

The curve for the hot-drawn (110°C), Series C isotactic polypropylene films in Fig. 2-10 (□ points) is intermediate between the curves for the cold-drawn, isotactic polypropylene film and the melt-spun, isotactic polypropylene fiber. The same rationale as that presented above for the melt-spun fibers can be used to explain the $b_\perp$ bow of this curve. A much smaller fraction of a'-axis oriented crystallites must be present at this temperature than at the melt-spinning temperature, however, since the $b_\perp$ bow for the hot-drawn films is much less than that for the melt-spun fibers. One can conclude from these observations that the amount of a'-axis orientation increases with draw temperature and decreases with extension.

The above discussion of the orientation behavior of isotactic polypropylene crystallites has been used to illustrate the power of the crystal orientation function and its representation in an orientation function triangle diagram. This system of representation can clearly illustrate the effects of environment on the orientation behavior of crystallites during deformation. A knowledge of crystallite orientation behavior is vital to understanding the effect of processing and fabrication conditions on the mechanical properties of commercial films.

3. Sonic Modulus: Determination of the Amorphous Orientation Function

a. Introduction. When a polycrystalline polymer film is oriented, the deformation ultimately leads to a rearrangement of the crystalline and amorphous regions of the film. To describe the morphological state of the deformed film completely, one must have a knowledge not only of the final state of orientation of the crystallites but of the final state of orientation of the amorphous region as well. Thus, a knowledge of the value of the amorphous orientation function is essential if the state of orientation of a deformed film is to be described.

Due to the less-ordered arrangement of the molecules in the amorphous region, an experimental determination of its orientation function becomes more elusive than that of the crystalline regions. In principle, the distribution of orientations of the amorphous chains could be determined from an analysis of the intensity distribution of the amorphous halo in the wide-angle x-ray diffraction pattern (35). However, the separation of the amorphous from the crystalline contributions to the pattern is very difficult experimentally. Similarly, in principle, infrared measurements might be used to determine the amorphous orientation function (35). However, characterization of the amorphous contribution to the infrared spectra is experimentally very difficult and the transition moment angle at the frequency used in the investigation must be known. The use of infrared measurements for the determination of molecular and orientation characteristics of polymers is treated more fully in a later section.

A study of the manner in which a sonic wave propagates through a polycrystalline polymeric medium can give considerable insight into the nature of the orientation of the two phases in the medium. To understand the rationale behind this measurement the models used to describe sound propagation must first be examined.

b. A Model for the System. The nature of sound propagation in uniaxially oriented fibers has been examined by Ward (36, 37) and Moseley (38). Ward considered a partially oriented fiber as an aggregate of units whose optical and elastic properties are those of a highly oriented fiber. The polymer is thus treated as a single-phase system whose properties are a function of the average distribution of the units that make up the system. The units are assumed to be ideally elastic materials, which possess transverse isotropy and whose properties are unchanged by the process of orientation. Using this model, Ward applied a continuum mechanics analysis to define the elastic constants for the system and then further evaluated these in terms of the orientation distribution of the units. In this way he was able to demonstrate, as a special case of the aggregate theory, the following relationship between the sonic modulus E, the intrinsic lateral (transverse) modulus of the perfectly oriented fiber E_t^0, and the average angle θ between the direction of sound propagation and the symmetry axis of the units

$$\frac{1}{E} = \frac{1 - \overline{\cos^2 \theta}}{E_t^0} \tag{2-27}$$

Moseley examined the sonic modulus of several stiff, oriented polymers whose glass-transition temperatures were very far above the room temperature at which they were measured. He found that in this frozen state, the sound propagation of the samples was independent of the degree of crystallinity. For this reason, he developed a theory of sound propagation that considered the fiber as a single-phase system. He then analyzed the molecular processes by which the sound was propagated through the system. Moseley reasoned that if sound is sent across an array of parallel molecules, the sonic energy is transmitted from one molecule to another by the stretching of intermolecular bonds (Fig. 2-12). On the other hand, if the sound is sent along the length of a bundle of parallel molecules, the sound is propagated principally by the stretching of chemical bonds in the backbone of the polymer chain. For partially oriented polymer molecules the molecular motion due to sound transmission is presumed to have right-angle components along and across the direction of the molecular axis. The magnitude of either of these two components is a function of the angle θ between the molecular axis and the direction of sound propagation. The only physically meaningful assumption about the method of addition of the two components is that any fiber deformation that occurs during sound transmission is the sum of the intramolecular and intermolecular deformations. Thus, he concluded that a series addition of the intra- and intermolecular force constants, weighted by an average orientation parameter, had to be used to represent the propagation of a sound pulse through a fiber. Assuming the force constants to be proportional to the corresponding sonic moduli, Moseley derived the following equation for the sonic modulus E:

$$\frac{1}{E} = \frac{1 - \overline{\cos^2 \theta}}{E_t^0} + \left(\frac{\overline{\cos^2 \theta}}{E_1^0}\right) \tag{2-28}$$

where E_t^0 is the intrinsic lateral (transverse) modulus of a fully oriented fiber (i.e., where all of the molecules are aligned parallel to each other and perpendicular to the direction of sound propagation) and E_1^0 is the intrinsic longitudinal modulus of a fully oriented fiber (i.e., where all of the molecules are aligned parallel to each other and parallel to the direction of sound propagation). Because of the high longitudinal sound velocity along the chain axis of the molecule, the second term on the right in eq. 2-28 was considered to be negligible. Thus, eq. 2-28 reduced to

$$\frac{1}{E} = \frac{1 - \overline{\cos^2 \theta}}{E_t^0} \tag{2-29}$$

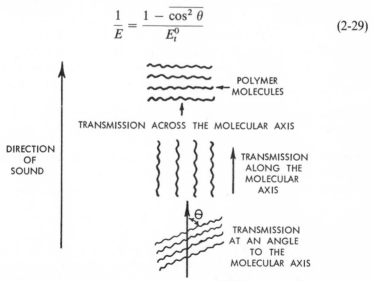

Fig. 2-12 Possible modes of sound transmission in polymers.

Equation 2-29 is identical to eq. 2-27. Thus, both authors, starting from different models, obtained the same equation to describe the mechanism of sound propagation in a uniaxially oriented, single-phase polymer system.

Polyethylene, nylon 6,10, Teflon (39), and polypropylene (25,27,40) all show changes in sonic modulus with percentage crystallinity when measured at room temperature. These materials do not behave as single-phase systems, but as two-phase systems when responding to a sonic pulse. Polypropylene behaves as a two-phase system even at temperatures far below its glass-transition temperature. To be applicable to such polymers, the sonic modulus theory must be extended to a two-phase model, and must take into account different intermolecular force constants for the crystalline and amorphous regions.

To study the orientation behavior of isotactic polypropylene, Samuels extended the sonic modulus equations to a two-phase system (25) by use of a mixing equation involving bulk compressibilities. For a homogeneous ideal mixture, both the density and bulk compressibility are additive properties. If isotactic polypropylene is considered an ideal mixture of amorphous and crystalline phases, then the mixing equation takes the form

$$K = \beta K_c + (1 - \beta)K_{am} \tag{2-30}$$

where K is the bulk compressibility of the mixture, K_c is the bulk compressibility of the crystal regions, K_{am} is the bulk compressibility of the amorphous regions, β is the fraction of crystalline material, and $(1 - \beta)$ is the fraction of noncrystalline material.

An equation of this form has been found to be valid for isotropic suspensions of solids in liquids (41) and for polyethylene over a range of crystallinities and temperatures (42). The bulk compressibility K is related to the bulk modulus B, Young's modulus E, and Poisson's ratio v, by the expression (43)

$$K = \frac{1}{B} = \frac{3(1 - 2v)}{E} \tag{2-31}$$

Waterman (44) found that $v = 0.33$ for isotactic polypropylene at room temperature. For this specific case, then,

$$K = \frac{1}{B} = \frac{1}{E} = \frac{1}{\rho C^2} \tag{2-32}$$

where the sonic modulus (Young's modulus) $E = \rho C^2$ for a long, thin, rodlike sample (45). Here ρ is the density and C the sonic velocity. Equation 2-29 is assumed to be valid for each phase. Then combination of eqs. 2-29, 2-30, and 2-32 yields the following expression for the measured sonic modulus

E_{or} of the oriented sample:

$$\frac{1}{E_{or}} = \left(\frac{\beta}{E_{t,c}^0}\right)(1 - \overline{\cos^2 \theta_c}) + \left(\frac{1 - \beta}{E_{t,am}^0}\right)(1 - \overline{\cos^2 \theta_{am}}) \qquad (2\text{-}33)$$

where the subscripts c and am stand for the crystalline and amorphous regions, respectively.

For an unoriented sample, $\overline{\cos^2 \theta} = 1/3$ and eq. 2-33 reduces to

$$\frac{3}{2E_u} = \frac{\beta}{E_{t,c}^0} + \frac{1 - \beta}{E_{t,am}^0} \qquad (2\text{-}34)$$

where E_u is the measured sonic modulus of the unoriented sample. Equation 2-34 predicts that the sonic modulus will vary in a predescribed way as a function of the fraction of crystals in the unoriented samples. Thus, by studying the change in the sonic modulus of unoriented samples with crystallinity, the intrinsic lateral modulus of the crystalline and of the amorphous regions of the polymer may be determined.

The orientation function f_x was defined as

$$f_x \equiv \frac{3 \overline{\cos^2 x} - 1}{2} \qquad (2\text{-}9)$$

where x represents the angle between the polymer-chain axis and a specified reference direction in the sample (here the stretch direction, which is also the direction of sound propagation). By applying this expression to eq. 2-33 and combining it with eq. 2-34, the following expression for the sonic modulus of oriented, isotactic polypropylene at room temperature is obtained:

$$\tfrac{3}{2}(\Delta E^{-1}) = \frac{\beta f_c}{E_{t,c}^0} + \frac{(1 - \beta)f_{am}}{E_{t,am}^0} \qquad (2\text{-}35)$$

where

$$(\Delta E^{-1}) = (E_u^{-1} - E_{or}^{-1})$$

and f_c and f_{am} are defined orientation functions for the crystal and amorphous phases, respectively. Thus, according to eq. 2-35, once the intrinsic lateral moduli of the crystalline and amorphous regions have been determined from measurements on unoriented samples, the amorphous orientation function f_{am} of any oriented sample can be determined from experimental values of the sonic modulus E_{or}, the fraction of crystals β, and the crystal orientation function f_c.*

* The derivation of eq. 2-35 is strictly valid only when Poisson's ratio v has a value of 0.333. Thus, eq. 2-35 is valid for isotactic polypropylene at room temperature. When a temperature other than room temperature is used to obtain sonic moduli of isotactic polypropylene, or when other polymers are studied whose Poisson's ratio is different from 0.333 at the measurement temperature, eq. 2-31 must be used to correct the theoretical derivation for the new Poisson's ratio.

c. Determination of the Sonic Modulus.   The sonic modulus for a long, thin, rodlike sample is given by the expression (45)

$$E = \rho C^2$$

where E is the sonic modulus (Young's modulus), ρ is the density, and C is the sonic velocity. Obviously, a long, thin, rodlike sample must be used to determine the sonic modulus. Fibers make ideal specimens but long, thin films can also be used. Film samples 1–5 mils thick, 1 mm wide, and 15–25 cm long are quite satisfactory. The density of portions of this film can be measured easily and rapidly in a density gradient column (46).

The sonic velocity (the velocity of a longitudinal wave in the material) is usually determined by measuring the transit time of a sound pulse between two tranducers coupled to a test specimen. The transit time of the pulse over a given length of specimen is inversely proportional to the propagation velocity in the sample. Figure 2-13 represents schematically the general features of an experimental apparatus for measuring the sonic velocity of a polycrystalline polymer specimen. The sample film strip is clamped at one end, passed over a pulley, and kept taut by a 10-g weight (for light tension) attached to the other end. This arrangement can be varied by replacing the pulley and weight with a clamp attached to a strain gauge and applying a force of 10 g to the strip in this manner. An optical bench is a good support for the test equipment. The transmitter and receiver (transducers) can then be moved along the optical bench and the distance between the probes can be read from a meter stick. The sound pulse is supplied by a pulse propagation meter. The KLH Series Four Pulse Propagation Meter, which produces longitudinal waves at a frequency of 10 kc/sec has been the standard workhorse in this field. The operator simply places the transducers a given distance apart in contact with the specimen and reads the transit time directly in microseconds (μsec) on the panel meter or the attached recorder. This is measured at several distances between probes and the results are plotted as in Fig. 2-14. The slope of the line is equal to $1/C$ (μsec/cm).

The Series Four Pulse Propagation Meter is no longer available from the

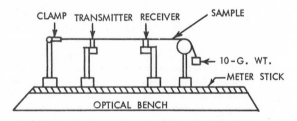

Fig. 2-13   Schematic representation of sonic-velocity apparatus.

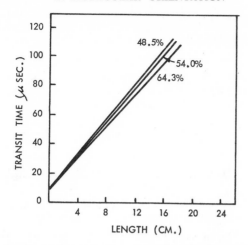

Fig. 2-14 Sonic-velocity determination of unoriented polypropylene films of different crystallinities.

KLH R & D Corp. Dr. Morgan, who designed the meter, is now the supplier (H. M. Morgan Co., Inc.) of the PPM-5 (Fig. 2-15). This is a solid-state version of the Series Four with some added conveniences. The apparatus is supplied with a bench, clamps, and transducer mounts. One of the transducers moves along the sample at a constant rate and the transit time is monitored directly as a function of the distance between the probes on an

Fig. 2-15 The Dynamic Modulus Tester PPM-5 (H. M. Morgan Co., Inc.).

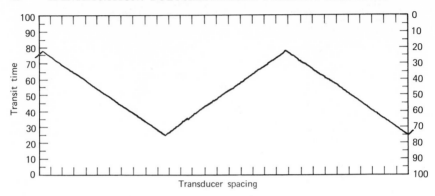

Fig. 2-16   Recorder trace from the Dynamic Modulus Tester PPM-5.

attached recorder. The moving probe cycles continuously. The output for two cycles is illustrated in Fig. 2-16. The operator simply calculates the sonic velocity from the reciprocal of the slope of the line plotted by the instrument. The reproducibility from cycle to cycle is about 1 %.

Once the sonic velocity has been measured, the sonic modulus is calculated directly from the known density and velocity values. The sonic velocity measurement is very rapid and requires but a few minutes. This makes it an attractive tool for orientation studies.

d. Application of the Sonic Modulus to Orientation Studies.   The sonic modulus is a mechanical measurement. It measures the Young's modulus of the specimen. The sonic pulse causes real displacements of the molecules from their equilibrium position as it passes through the sample.

In the two-phase system, the intermolecular force constants for each phase are different. Since the intrinsic lateral modulus is a mechanical measurement, the unoriented samples will have nonzero sonic modulus values. In fact, the measured sonic modulus value will represent the additive properties of each individual phase weighted by the amount of each phase present. Equation 2-34 describes the law of addition of the properties of the two phases. Table 2-1 lists the variation in sonic modulus with β for a series of unoriented, compression-molded, isotactic polypropylene films (25) (see Fig. 2-14). The intrinsic lateral modulus of the crystal $E_{t,c}^0$ and of the amorphous region $E_{t,am}^0$ can be calculated from these experimental values of the sonic modulus and the fraction of crystals β by using eq. 2-34.

The intrinsic lateral modulus of the crystal $E_{t,c}^0$ is the intermolecular Young's modulus of the chains. If the polymer chains are in a folded, single crystal with the chain axes all aligned along a vertical axis Z and in a given crystallographic plane [such as the (110) for polyethylene], the intrinsic lateral modulus may be visualized as the force per molecule required to

Table 2-1. Sonic Modulus (E_u), Crystal Fraction (β), and Density of Unoriented Compression-Molded Isotactic Polypropylene Films (25)

Sample	$E_u \times 10^{-10}$, dyne/cm^2	Density, g/cm^3	β
1	2.27	0.8875	0.410
2	2.48	0.8936	0.485
3	2.63	0.8980	0.540
4	3.01	0.9061	0.643

separate the planes of molecules [(110) faces] a given distance by applying the force in the direction perpendicular to the Z axis and normal to and away from the crystal face. The intrinsic lateral modulus of the isotactic polypropylene crystal, calculated from the experimental values of E_u and β determined at room temperature, is:

$$E_{t,c}^0 = 3.96 \pm 0.09 \times 10^{10} \text{ dyne/cm}^2$$

Another method of measuring the intrinsic lateral modulus of the crystal is to measure the force required to cause a given displacement of a known plane in the crystal lattice. The displacement of the crystal lattice is measured as a change in the Bragg-angle spacing for the given lattice plane in the wide-angle x-ray diffraction pattern. This technique was used by Sakurada, Ito, and Nakamae (47) to determine the intrinsic lateral modulus of the isotactic polypropylene crystal. They obtained a value of $E_{t,c}^0 = 2.8 - 3.1 \times 10^{10}$ dyne/cm^2 for isotactic polypropylene. This is rather good agreement with the value obtained from the sonic modulus method, especially when one realizes that different investigators (48,49), each using the same stress–x-ray diffraction technique, obtained values of $E_{t,c}^0$ for polyethylene ranging from 2.2×10^{10} to 4.2×10^{10} dyne/cm^2.

The intrinsic sonic modulus of the amorphous region is defined as the transverse Young's modulus that the amorphous chains would have in a perfectly oriented fiber. The lateral forces between amorphous chains in this system would be expected to be lower than those in the crystal lattice. The calculated intrinsic lateral modulus of the amorphous region of isotactic polypropylene is

$$E_{t,am}^0 = 1.06 \pm 0.01 \times 10^{10} \text{ dyne/cm}^2$$

which is, as predicted, lower than the value of the intrinsic lateral modulus obtained for the crystal.

Once the material constants for the polymer have been determined from sonic modulus studies on unoriented specimens, the sonic modulus of oriented specimens can be examined. The sonic modulus of an oriented film

Table 2-2. Sonic Modulus (E), Crystal Fraction (β), and Density of Cast and Annealed Series C Isotactic Polypropylene Films (25)

Sample	Elongation, %	$E \times 10^{-10}$, dyne/cm^2	Density, g/cm^3	β
1	0	2.86	0.9034	0.608
2	50	3.09	0.9052	0.630
3	100	3.59	0.9056	0.636
4	200	5.07	0.9055	0.636
5	300	6.19	0.9060	0.643
6	400	6.55	0.9052	0.630

is a measure of the total orientation of the sample. It is a function of the orientation of both the crystalline and amorphous regions as well as that of the amount of each phase present in the sample. The rule for addition of the properties of each of the phases is defined by eq. 2-35. The dependence of the sonic modulus E_{or} on elongation for Series C isotactic polypropylene films [cast and drawn at 110°C (25)] is illustrated in Table 2-2 and Fig. 2-17. Since the intrinsic lateral moduli are now established, eq. 2-35 permits calculation of the amorphous orientation function f_{am} for any oriented,

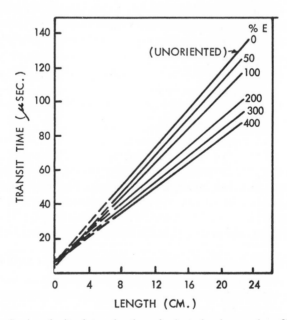

Fig. 2-17   Sonic-velocity determination of oriented polypropylene films.

isotactic polypropylene sample from experimental values of the sonic modulus (E_{or}), density (β), and the crystal orientation function (f_c) determined by x-ray diffraction. Values of f_{am} determined for the oriented, isotactic polypropylene samples are listed in Table 2-3.

The orientation of the crystalline and amorphous phases can be determined

Table 2-3. Birefringence (Δ_T) and Orientation Function (f) Data for Series C Isotactic Polypropylene Films (25)

Sample, Cast and Annealed	Elongation, %	Birefringence, $\Delta_T \times 10^3$	X-ray f_c	$-f_\beta$	Sonic Modulus, f_{am}
2	50	5.759	0.3472	0.2485	−0.0951
3	100	10.19	0.4790	0.2898	+0.0232
4	200	19.64	0.6805	0.3775	+0.2849
5	300	22.25	0.7728	0.4022	+0.3877
6	400	25.30	0.8016	0.3988	+0.4321

quantitatively by a combination of density, sonic modulus, and x-ray diffraction methods. With the sample thus fingerprinted, a potential exists for relating the observed mechanical properties of these samples to their morphological characteristics (see Chapter 4). Similarly, with this information, the utility of other techniques may be extended and new molecular information about the polymers obtained. For example, by combining this information with birefringence measurements a rapid method of obtaining the orientation functions can be developed and at the same time a method for determining the intrinsic birefringence of each of the phases is obtained.

4. Birefringence: Determination of the Total Molecular Orientation

a. **Introduction.** Birefringence is a measure of the total molecular orientation of a system. It is defined as the difference in the principal refractive index parallel $n_{\parallel}$ and perpendicular $n_{\perp}$ to the stretch direction for a uniaxially oriented specimen. The refractive index, in turn, is a measure of the velocity of light in the medium and is related to the polarizability of the chains. Thus, the final measured birefringence is a function of contributions from the polarizabilities of all of the molecular units in the sample.

The velocity of light in the crystalline region will be different from that in the amorphous region. Since the molecules are optically anisotropic, there is a difference in the principal polarizability parallel $P_{\parallel}$ and perpendicular $P_{\perp}$ to the chain axis. If all of the anisotropic molecules are randomly distributed in the specimen (unoriented sample), the average contribution to the refractive index will be the same in all directions and the birefringence, $\Delta_T =$ $(n_{\parallel} - n_{\perp})$, will be zero (i.e., the specimen is optically isotropic). If all of the molecules are perfectly aligned parallel to the stretch direction, the birefringence will be a combination of the ideal polarizability anisotropy of each region; that is, each region will have the birefringence of a fully oriented fiber of that phase alone. The measured birefringence of oriented sample films will have intermediate values between these extremes, depending on the orientation and amount of each phase present in the specimen.

For a uniaxially oriented, polycrystalline polymer (assuming a two-phase model of crystalline and amorphous regions) the measured birefringence Δ_T may be defined as (9)

$$\Delta_T = \beta\Delta_c^0 f_c + (1 - \beta)\Delta_{am}^0 f_{am} \tag{2-36}$$

Here Δ_c^0 is the intrinsic birefringence of the perfectly oriented crystal and Δ_{am}^0 is the intrinsic birefringence of a perfectly oriented amorphous region (i.e., the birefringence the amorphous chains would have in a perfectly oriented amorphous fiber). Equation 2-36 neglects an added term, the form birefringence (50), which takes into account the effect of the shape of the crystallite in the medium. The effect has been found to contribute 5–10% of the total birefringence, depending on orientation, to the measured birefringence of polyethylene (51). The form birefringence, is negligible for the isotactic polypropylene samples.

Birefringence should be an excellent property to use for the study of orientation in a polycrystalline polymer. Once the intrinsic birefringences have been obtained, the amorphous orientation function could be determined, for instance, by a combination of density, x-ray diffraction, and birefringence measurements. The problem in applying this method to a wide spectrum of polymers has been the difficulty in obtaining intrinsic birefringence values of the polymer. Unlike the sonic modulus, for which the intrinsic lateral moduli can be obtained from measurements on unoriented samples of different crystallinities, the birefringence of an unoriented specimen is zero. Thus, one cannot measure directly the intrinsic birefringences of a polymer by examining the birefringence of unoriented specimens.

This has been a major handicap in the use of birefringence measurements for orientation studies on polycrystalline polymers. To date, birefringence has been used to determine the amorphous orientation function of only a few polymers, such as polyethylene (52), polypropylene (25–28), polyethylene terephthalate (30–32), and hydroxypropylcellulose (33). For the polyethylene studies, the birefringence of the crystalline region was assumed to be the same as that obtained from single crystals of low molecular weight (C_{36}) paraffin homologs (53), whereas the intrinsic birefringence of the amorphous region had to be calculated theoretically from bond-polarizability values. By use of these values, in combination with x-ray diffraction and density measurements, the birefringence contribution of the crystalline and amorphous regions was calculated. The amorphous orientation function of the oriented samples was then estimated from these results.

A study of the birefringence of oriented, isotactic polypropylene films was carried out to test the validity of amorphous orientation functions obtained from sonic modulus measurements. In the process of doing this, Samuels developed a method for obtaining the intrinsic birefringence of the crystalline and amorphous regions from experimental data (35), and hence the practical utility of the birefringence measurement was extended. Before considering this work in more detail, a description of the experimental determination of birefringence will be given.

b. Determination of Birefringence. Light propagates by wave motion through a medium. The standard relationship between frequency v, wavelength λ, and velocity c for wave motion is

$$c = v\lambda \qquad (2\text{-}37)$$

The frequency of a beam of monochromatic light never changes, even if the light enters an entirely different material. The wavelength and velocity of the same light beam do change if it enters a different material. The index of refraction n of a particular material is defined as

$$n = \frac{c_0}{c_m} = \frac{\lambda_0}{\lambda_m} \qquad (2\text{-}38)$$

The subscripts 0 and m stand for vacuum and the medium, respectively. The velocity in air is almost the same as that in vacuum and hence the refractive index of air may be assumed equal to 1.0 in most cases (the actual value is $n = 1.0003$). A light wave is slowed down as it goes from vacuum to any material and hence the refractive index is greater than 1.0 for all materials.

The mechanism of light propagation through a medium is related to the polarizability of the molecules in the medium.

$$P = N\alpha \qquad (2\text{-}39)$$

where P is the polarizability per unit volume, N is the number of molecules, and α is the polarizability of a molecule. The polarizability α, in turn, is a consequence of the electric field of the incoming light wave E inducing a dipole moment μ in the electrons of the molecule. The polarizability α is the ratio of the induced dipole

$$\alpha = \frac{\mu}{E} \qquad (2\text{-}40)$$

moment to the applied field and therefore is essentially a measure of the mobility of the electrons in the molecules.

The origin of the refractive index of a material is the polarizability of the molecules. The refractive index can be related to the principal polarizability of the molecule by the Lorenz–Lorenz equation

$$\frac{n^2 - 1}{n^2 + 2} = \frac{4}{3}\pi P \tag{2-41}$$

Most molecules are optically anisotropic (have different polarizabilities along and across the molecule). When the anisotropic molecules are randomly distributed throughout the medium, the average polarizability in any direction is the same and a single refractive index will characterize the system.

When a polymer film is uniaxially oriented, the chain axis of the molecules orients with respect to the deformation axis and the molecules have cylindrical symmetry around the deformation axis. Since the optically anisotropic molecules now have a preferred direction with respect to the deformation axis of the film, the film now manifests anisotropic optical properties. For this system, the principal refractive index for the light vibrating in the deformation (stretch) direction is different from that for light having its electrical vibration in the plane perpendicular to the deformation direction. The birefringence is used to characterize such an optically anisotropic system. For a uniaxial system, the birefringence Δ_T is defined as the difference between the principal refractive index in the stretch direction $n_{\parallel}$ and the principal refractive index perpendicular to the stretch direction $n_{\perp}$:

$$\Delta_T = n_{\parallel} - n_{\perp} \tag{2-42}$$

To determine Δ_T for a sample, the individual principal refractive indices of the sample parallel and perpendicular to the deformation axis must be determined and then subtracted, or else the difference in the two principal refractive indices must be measured directly.

The direct measurement of the individual principal refractive indices of the sample is a tedious and difficult procedure, fraught with possible errors, and very time consuming. Monochromatic light is used because different frequencies of light will have different velocities in the sample. The sample to be measured is split into several subsamples, which must be placed in a series of immersion liquids of different refractive index. They are then examined under a polarizing microscope. The plane of the incident light is adjusted parallel to the stretch direction of the subsamples and the refractive index $n_{\parallel}$ is determined by the Becke method from examination of the series. The plane of the incident light is then adjusted perpendicular to the stretch direction of the subsamples and the refractive index $n_{\perp}$ is determined by another examination of the series. The observer, when using the Becke method, looks for a thin band of light at the edge of the sample. This band of light is called the Becke line. Under the microscope, the narrow bright band moves toward the medium of higher refractive index if the focus of the microscope is raised, and toward the medium of lower refractive index if the focus is lowered. With the use of monochromatic light and controlled temperature an accuracy of ± 0.001 in the refractive index may be reached by successive approximation when proper selection of immersion media is made (54).

A more rapid and popular method for determining birefringence is the use of a compensator to determine the phase difference (retardation) between two mutually perpendicular, plane-polarized wave motions emerging from the sample. Since the sample is anisotropic, the velocity of the wave passing through the sample parallel to the stretch direction will be different from the velocity of the wave oriented perpendicular to the stretch direction. This velocity difference between the two waves causes a phase difference in the emerging rays. The phase difference, called the retardation, is proportional to the birefringence. By measuring the retardation, the birefringence can be determined from the relation

$$\Delta_T = \frac{\lambda R}{t} = \frac{\Gamma_\lambda}{t} \tag{2-43}$$

where λ is the wavelength of light (nm), R is the retardation (phase difference in wave numbers—dimensionless), Γ_λ is the retardation at a specified λ (nm), and t is the thickness of the sample (nm).

Experimentally, the retardation of the sample is determined by matching the sample's retardation with an equal and opposite retardation using a standard calibrated crystal. The standard compensating crystal with its vernier and holder is called a compensator. Several different compensator designs are available. They all work on the same physical principle but differ in their mechanical arrangement. The most common are the Babinet compensator, which is made up of a quartz wedge whose thickness can be varied by sliding different regions of the wedge into the light path; the Berek compensator, which varies the thickness and hence the retardation by a rotation of the calibrated crystal; and the Ehringhaus rotatable compensator with a crystal combination plate. The author uses a Zeiss polarizing microscope with Ehringhaus compensators. The advantage of the Ehringhaus compensator is that the usual zero reading with the sample removed from the beam is eliminated. Instead, due to the properties of the combination plate, the zero-order, black extinction line is obtained by rotating the crystal plate first in one direction and taking a reading and then in the opposite direction for a second reading. The average value of the two vernier readings gives the correct retardation of the sample in degrees of inclination of the plate. The result can be converted to the desired phase difference in nanometers (millimicrons) by calculation. Actually, the Ehringhaus compensator comes with a table of phase differences for given inclination angles so that the retardation in nanometers can be obtained rapidly.

The experimental arrangement of the sample and optical components in the microscope is shown in Fig. 2-18. The light beam passes through a

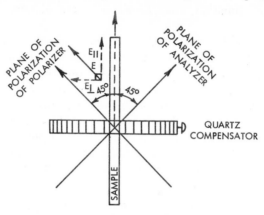

Fig. 2-18   Arrangement of the optical system for a retardation measurement. The incident light beam is directed perpendicular to the surface of the page.

polarizer that has its axis at 45° to the stretch direction of the sample film. The incident beam E can be resolved into a component $E_{\parallel}$, parallel to the stretch direction, and a component $E_{\perp}$, perpendicular to the stretch direction. The phase difference between these two components in the emerging wave depends on the birefringence of the sample. The emerging ray then enters the compensator whose direction of retardation is opposite to that of the sample film. Finally, an analyzer is oriented perpendicular to the plane of polarization of the incident light and 45° to the stretch direction of the sample film. Initially, white light is used to find the zero-order, black extinction fringe in the compensator. This fringe is set on the cross hairs of the microscope and then a monochromatic interference filter is placed in the beam. This sharpens the extinction line under the cross hairs, which are then recentered. White light must be used initially because the fringes of all orders are black under monochromatic radiation and hence white light must be used to identify the zero-order fringe.

The author uses a green line ($\lambda = 546$ nm) interference filter and has used both quartz and calcite Ehringhaus compensators. With the quartz compensator, one has an available measuring range of about seven orders, while with the calcite compensator the available measuring range is about 133 orders. The time required to obtain a retardation measurement on a transparent film is only a few minutes. Thus, the birefringence measurement is very rapid.

To calculate the birefringence of the sample, the measured retardation Γ_λ (nm) and the measured thickness (nm) (1 mm $= 10^6$ nm) are substituted into eq. 2-43.

A method has been developed for determining the retardation from the light intensity transmitted through a system similar to Fig. 2-18 but without the compensator (55). This method is most suitable for dynamic birefringence measurements and will not be considered in detail here.

c. Application of Birefringence to Orientation Studies.   Birefringence is used routinely in many industrial film and fiber plants as a measure of the average orientation of the sample. This reasoning ignores the complex nature of birefringence and treats the sample as if it were a single, average-phase system. This single, average-phase, molecular orientation concept pervades much of the literature and has led some investigators (37,56,57) to attempt, on this basis, to correlate birefringence with other properties of the polymer. Other investigators have realized the complex nature of birefringence, but due to the lack of availability of one or both of the intrinsic birefringence values, they could not make use of birefringence data (23,58). A knowledge of the intrinsic birefringence of the crystal and amorphous regions of a polycrystalline polymer is essential if maximum information is to be obtained from the measured birefringence of a sample.

The intrinsic birefringence of the crystal and amorphous regions of a polymer can be obtained if the fraction of crystals and the crystal and amorphous orientation functions are known for a series of oriented samples. Samuels has shown that eq. 2-36 can be rewritten in the following form (25):

$$\frac{\Delta_T}{\beta f_c} = \Delta_c^0 + \Delta_{am}^0 \left(\frac{1-\beta}{\beta}\right)\left(\frac{f_{am}}{f_c}\right) \tag{2-44}$$

Equation 2-44 predicts a straight line relationship between $\Delta_T/\beta f_c$ and $[(1-\beta)f_{am}/\beta f_c]$. Experimental values of Δ_c^0 and Δ_{am}^0 can then be obtained directly from the intercept and slope, respectively, of the resulting line. Valid experimental values of Δ_T, β, f_c and f_{am} are required to satisfy this equation. Thus this equation not only offers a direct experimental route to the intrinsic birefringence values, but also offers the opportunity of testing the validity of the f_{am} values obtained experimentally from sonic modulus measurements.

Samuels' two-phase sonic modulus theory predicts that eq. 2-35 is valid irrespective of the manner in which the experimental samples used are prepared. f_{am} values were obtained from sonic modulus measurements on the Series C isotactic polypropylene films (Table 2-3) and from two other film series (A and B; in Table 3-1) discussed in detail in Chapter 3. Together the Series A, B, and C samples represent films that differ in molecular weight of the polymer, have as much as a 10% difference in crystallinity, have been drawn from 12 to 815% extension at two different temperatures, and vary from low to very high orientation. Thus, these 18 films constitute an excellent series for severely testing the validity of the two-phase sonic modulus theory.

A plot of $\Delta_T/\beta f_c$ against $(1 - \beta) f_{am}/\beta f_c$ for the 18 oriented series A, B, and C samples (Fig. 2-19), shows the straight-line behavior predicted by eq. 2-44.

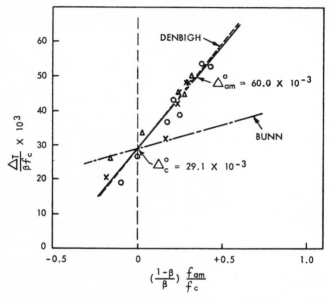

Fig. 2-19 Determination of the intrinsic birefringences of isotactic polypropylene: (O) Series A; ($\times$) Series B; ($\triangle$) Series C (27).

Each point on this plot is a combination of a birefringence, a density, an x-ray diffraction, and a sonic modulus measurement. The ratio f_{am}/f_c should be very sensitive to errors in f_{am}. The results indicate that eq. 2-35 is independent of the method of preparation of the oriented films and is dependent only on the resulting morphological state of orientation.

The intrinsic birefringence of the crystalline ($\Delta_c^0 = 29.1 \times 10^{-3}$) and non-crystalline ($\Delta_{am}^0 = 60.0 \times 10^{-3}$) regions of isotactic polypropylene were calculated from the intercept and slope of the line in Fig. 2-19. These data represent the first time intrinsic birefringences of a polycrystalline polymer had been obtained from measurements made directly on the polymer (25,27).

The intrinsic birefringence of the crystal is found to be smaller than that of the amorphous region of isotactic polypropylene. The smaller value of Δ_c^0 is due to the ordered lattice arrangement of neighboring molecules in the crystal. A bond within the molecule experiences not only the electromagnetic field of the incident light but also the polarization field of the surrounding molecules. This internal field is more anisotropic in the ordered lattice of the crystal than in the disordered amorphous phase. Consequently, there are different effective polarizabilities in the two phases. Both Stein (59) and Vuks (60) have derived theories that satisfactorily account for internal field effects in the polyethylene crystal. They each concluded that, due to internal field effects, the intrinsic birefringence of the crystal will be lower than that of the amorphous material. Experimental results with isotactic polypropylene are consistent with their theoretical conclusion. For this reason, all theoretical considerations of the data must concentrate on the noncrystalline region.

Infrared measurements on isotactic polypropylene have shown that the configuration of the molecules in the noncrystalline region is a regular helix with three monomer units in the statistical repeat unit (3-mer helix) (61,62). This is the same molecular conformation that is found for the repeat unit in the isotactic polypropylene crystal (16). The intrinsic birefringence can be calculated for this model, by using $(P_\parallel - P_\perp)_{3\text{-mer}}$, the principle polarizability difference values calculated by Stein (64). He calculated one $(P_\parallel - P_\perp)_{3\text{-mer}}$ value from the bond polarizabilities obtained by Denbigh (64) from gas-phase measurements (no internal field effects) and another using the bond polarizabilities Bunn (53) obtained from crystals (which have an internal field). Theoretically only the $(P_\parallel - P_\perp)_{3\text{-mer}}$ calculated from Denbigh's bond polarizabilities should be valid. The intrinsic birefringence of the 3-mer helix in the noncrystalline region is calculated from the expression (25):

$$\Delta_{am}^0 = \frac{2}{9}\pi\left[\frac{(\bar{n}^2 + 2)^2}{\bar{n}}\right]N_{3\text{-mer}}(P_\parallel - P_\perp)_{3\text{-mer}}$$

where $\bar{n}$ is the average refractive index of the polymer and $N_{3\text{-mer}}$ is (amorphous density/molecular weight of 3-mer unit) times Avogadro's number. From Denbigh's bond polarizabilities for $(P_\parallel - P_\perp)_{3\text{-mer}}$, the calculated intrinsic birefringence is $\Delta_{am}^0 = 61.5 \times 10^{-3}$. This is in excellent agreement with the value, $\Delta_{am}^0 = 60.0 \times 10^{-3}$, obtained experimentally. The calculated value of Δ_{am}^0 obtained with Bunn's bond polarizability values is 13.8×10^{-3}, which, as expected, does not agree with the experimental results. The significance of these results is shown in Fig. 2-19 in which the theoretical slopes are compared with the experimental data. The excellent agreement between the theoretical and experimental values of Δ_{am}^0 indicates that the 3-mer helix model of the isotactic polypropylene molecule in the noncrystalline region, as suggested by infrared evidence, is valid.

A further test of the two-phase sonic modulus theory is suggested by eqs. 2-35 and 2-36. They predict that a plot of the measured birefringence Δ_T against the measured sonic modulus in the form $3/2(\Delta E^{-1})$ will yield a straight line with a zero intercept and a positive slope. The data from the Series A, B, and C films when plotted in the suggested manner are found to behave as predicted (Fig. 2-20).

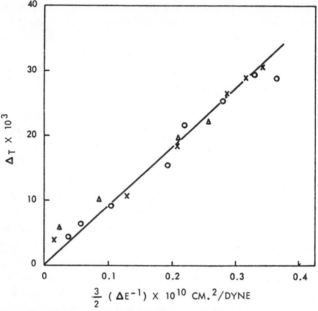

Fig. 2-20 Relation between the birefringence and the sonic modulus of isotactic polypropylene films: (O) Series A; ($\times$) Series B; ($\triangle$) Series C (27).

In all respects the f_{am} values calculated from the two-phase sonic modulus theory (eq. 2-35) have satisfied the severe requirements imposed by the complex morphological states and extreme processing conditions of the Series A, B, and C films. It may thus be stated with confidence that the two-phase sonic modulus theory is verified and the f_{am} values can be used to describe molecular orientation quantitatively in the noncrystalline regions of the films.

The above results indicate the potential importance of being able to fingerprint the morphological features of a polycrystalline polymer. By characterizing the morphological features of the isotactic polypropylene samples, the intrinsic birefringences of the polymer could be determined. With this new information, other characteristics of the polymer can be elucidated. For example, the contribution of the separate phases to the total measured birefringence can be calculated. These are plotted for the Series C isotactic polypropylene film samples in Fig. 2-21 where $\Delta_c = \Delta_c^0 f_c$ and $\Delta_{am} = \Delta_{am}^0 f_{am}$. The negative amorphous contribution at low elongations in Fig. 2-21 indicates that the amorphous chains are oriented toward the perpen-

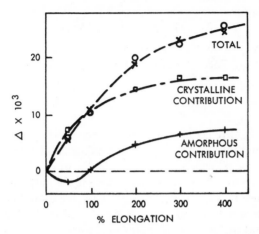

Fig. 2-21 Crystalline and amorphous contributions to the birefringence of deformed Series C isotactic polypropylene films: (O) $\Delta = \Delta_T$ (experimental); ($\times$) $\Delta = \Delta_T$ (calculated); ($\square$) $\Delta = \beta \Delta_c$ (calculated); ($+$) $\Delta = (1 - \beta) \Delta_{am}$ (calculated) (25).

dicular to the direction of stretch. As elongation of the film continues, the amorphous chains turn into the stretch direction and the amorphous birefringence becomes positive. The crystalline-orientation contribution to birefringence is fairly high even at low elongations. The low measured birefringence at low elongations is obtained even with the high crystalline orientation because the negative effect of the amorphous birefringence decreases the experimental value. Similar results were observed from orientation studies on low-density polyethylene (52). Low-density polyethylene is about 50% crystalline and structurally similar to polypropylene; therefore, the similar behavior of their crystalline and amorphous regions is not surprising.

These results illustrate the danger of treating birefringence as a measure of the average orientation of the sample. The measured birefringence of the low-elongation samples of isotactic polypropylene films indicates little orientation of the system when, in fact, there is considerable orientation of the crystalline and amorphous regions of the samples. The birefringence value of these films may lead to confusion when it is compared with other properties of the films, such as mechanical properties, if the investigator assumes the films have little molecular orientation. On the other hand, a comprehensive morphological interpretation of the structure of the film could make apparently confusing data—obtained from samples with the same measured birefringence but different mechanical properties—seem obvious.

Once the crystalline and amorphous intrinsic birefringences have been obtained, a very rapid method can be developed for determining the crystalline and amorphous orientation functions of the polymer. The equation for birefringence is

$$\Delta_T = \beta \Delta_c^0 f_c + (1 - \beta)\Delta_{am}^0 f_{am} \qquad (2\text{-}36)$$

where Δ_c^0 and Δ_{am}^0 are now known constants and Δ_T can be determined in a matter of minutes. The sonic modulus equation for an oriented polymer film is:

$$\tfrac{3}{2}(\Delta E^{-1}) = \frac{\beta f_c}{E_{t,c}^0} + \frac{(1 - \beta)f_{am}}{E_{t,am}^0} \qquad (2\text{-}35)$$

where $E_{t,c}^0$ and $E_{t,am}^0$ are now known constants and (ΔE^{-1}) can be determined in a matter of minutes. The fraction of crystals β can be determined from a density measurement, which takes a few hours to equilibrate in a density gradient column, but requires not more than a few minutes of operator time. Substituting these measured results from the sonic modulus, birefringence, and density determinations into eqs. 2-36 and 2-35 and solving them simultaneously leads to the desired values for the orientation functions f_c and f_{am}.

As discussed earlier, birefringence is a result of the polarizability of the anisotropic units which make up the polymer chain. It should, therefore, be possible to obtain some information about the polarizability of the monomer units within the chain from birefringence studies on polymers. Since, in general, due to internal field effects, it is impossible to calculate the anisotropic polarizability of bonds by using, as point of departure, the optical properties of the substance in the crystalline state (65), Δ_{am}^0 and not Δ_c^0 should be used in any attempt to calculate monomer polarizability differences.

The principal polarizability difference per monomer unit $(a_{\parallel} - a_{\perp})_{mer}$ can be calculated for isotactic polypropylene from the expression (52)

$$(a_{\parallel} - a_{\perp})_{mer} = \frac{\Delta_{am}^0}{(2/9)\pi[(\bar{n}^2 + 2)^2/\bar{n}]N_{mer}} = +5.79 \times 10^{-25} \text{ cm}^3 \qquad (2\text{-}45)$$

where $\Delta_{am}^0 = 60.0 \times 10^{-3}$; $\bar{n}$ is the average refractive index of the polymer $= 1.51$; and $N_{mer} = $ (amorphous density/molecular weight mer unit) $\times$ Avogadro's number. Here amorphous density $= 0.858$; molecular weight mer unit $= 42$; and Avogadro's number $= 6.02 \times 10^{23}$. The measured value of $(a_{\parallel} - a_{\perp})_{mer}$ is a characteristic property of the polymer molecule and hence should be the same whether obtained from measurement of birefringence in the bulk polymer or from flow birefringence measurements of polymer solutions. Tsvetkov (66) has tabulated values for the anisotropy per monomer unit calculated from flow birefringence data for numerous polymers. The value of $(a_{\parallel} - a_{\perp})$ reported for polypropylene is $+3.5 \times 10^{-25}$ cm^3. The value calculated from Δ_{am}^0 was $+5.8 \times 10^{-25}$ cm^3. The agreement between measurements in the solid state and in dilute solution is quite striking, especially since the $(a_{\parallel} - a_{\perp})$ values for polymers are known to range from -35×10^{-25} cm^3 for poly-p-chlorostyrene to $+21 \times 10^{-25}$ cm^3 for ethylcellulose.

The birefringence measurement is thus a rapid and powerful tool for the study of morphological characteristics of deformed polycrystalline polymers. Since birefringence is a measure of the total molecular orientation of the two-phase system, its examination in conjunction with other physical measurements (sonic modulus, x-ray diffraction, density, etc.) yields considerable insight into the molecular characteristics of the bulk polymer. The major problem in its use is a knowledge of the intrinsic birefringences of the polymer system to be investigated. One possible route to the experimental evaluation of these morphological constants has been described above. The principle for their evaluation is a general one, however, and any combination of physical measurements that yields values for β, f_{am}, and f_c can be used in conjunction with birefringence to determine Δ_c^0 and Δ_{am}^0. Once these have been evaluated for a given polymer, the birefringence measurement can be used as an important method for the comprehensive, quantitative evaluation of the morphology of the deformed polymer.

5. Infrared Dichroism: Molecular Structure and Orientation Parameters

a. Introduction. So far, the primary emphasis in this chapter has been on techniques for the quantitative determination of molecular orientation in polycrystalline polymers. As a secondary consequence of these methods, certain molecular structure parameters such as the intrinsic birefringences and the intrinsic lateral moduli of the system have become available. In this section, there will be a turnabout, and the quantitative determination of a molecular structure parameter—the transition moment angle—will be the principal quantity of the measurement, while the molecular orientation results will be a secondary consequence of the overall technique. For this reason, a general knowledge of the infrared absorption process and of the theoretical approach to structural infrared studies is necessary for comprehension of the importance of the transition moment angle determination. A number of excellent books (67,68) and reviews (69,70) are available on this subject, and it is not within the scope of this chapter to give a detailed account of infrared structure analysis. An attempt will be made, however, to outline the general theory of infrared spectroscopy to orient the reader for the discussions on molecular structure and orientation parameters which follow.

Mention should be made at the outset, however, that infrared dichroism is an important quantitative technique for the determination of crystalline and amorphous orientation functions. The nature of the measurement holds out the promise of being able to measure rapidly and quantitatively the crystalline and amorphous orientation functions separately. Since this method, like the previous techniques, is nondestructive, it has a potentially important role in the "*in vivo*" study of deformation processes in polymers. The interdependence of the two important facets of infrared dichroism, molecular structure and orientation parameters, is the subject of this section.

b. General Infrared Theory. Absorption in the infrared region of the spectrum is a function of the internal energy of the molecule. Molecules absorb energy in this region according to quantum rules and, hence, a continuous spectrum of energy absorption is not observed; instead, energy is absorbed only at discrete frequencies. When absorption occurs in this radiant energy region, it changes the ground-state rotational and vibrational energy levels of the molecule and this produces rotational and vibrational motion of the atoms constituting the molecule. Since the rotational energy changes are smaller than the vibrational energy changes, the effect of the rotational modes of motion on the observed spectra is to act as a sort of background which influences the shape of the vibrational energy bands. Hence, unless one is examining the spectra under high resolution to resolve the rotational energy absorption, the observed spectra can be considered as a measure of the vibrational modes of energy absorption of the molecules, that is, a vibrational spectrum.

The simplest model for representing the vibratory motion of the atoms resulting from the process of energy absorption in a molecule is the motion of a group of balls (atoms) of different masses held together by springs (chemical bonds). Application of a force to displace and then release a ball would correspond to the absorption of infrared energy. Upon being released the ball is set into vibratory motion. The frequency of the oscillation and the amount of displacement of the ball is a function of the spring (force) constant k, the mass of the ball m, and the original applied force F. For the same initial displacement, the motion of the ball will be changed by substituting another spring for the first or by changing the mass of the ball. The force required to stretch a spring a short distance d is given by Hooke's law

$$F = kd \qquad (2\text{-}46)$$

Thus, for a given applied force the displacement of the ball will be a function of the spring (force) constant.

In a simple diatomic molecule (I–II), the only vibration that can occur is a periodic stretching along the I–II bond. Stretching vibrations resemble the oscillations of two bodies connected by a spring and for this case Hooke's law, eq. 2-46, is applicable as a first approximation. The vibrational frequency v (cm^{-1}) for stretching the I–II bond is given by the expression

$$v = \left(\frac{1}{2\pi c}\right)\left(\frac{f}{\mu}\right)^{1/2} \qquad (2\text{-}47)$$

where c is the velocity of light, f is the force constant of the bond, and μ is the reduced mass of the system defined as

$$\mu = \frac{m_{\mathrm{I}} m_{\mathrm{II}}}{m_{\mathrm{I}} + m_{\mathrm{II}}} \qquad (2\text{-}48)$$

where m_{I} and m_{II} are the individual masses of atoms I and II.

In actual practice, one is generally interested in molecules composed of numerous atoms having bonds of differing strength. Thus, the motion of not two balls connected by a spring, but, instead, many balls and springs, is involved when energy is absorbed. The motion of a particular group of balls may predominate in a given applied-force (absorption-frequency) region, however, depending on the masses and spring constants involved. For instance, if a given ball is displaced and the adjoining balls have large masses and spring constants, the motion will be essentially isolated to the originally displaced ball because the displacement of the adjoining balls will be negligible.

Thus, infrared absorption by the molecule changes the vibrational energy level of the system. The problem is to correlate this energy change with the structure and force constants of the molecule. For a harmonic oscillator approximation, this problem can be treated as a classical analysis of a vibrating system. The problem is then reduced to a determination of the classical vibration frequencies associated with the normal modes of oscillation of the system. The characteristic of a normal mode is that all nuclei in the molecule move with simple harmonic motion of the same frequency. Determination of the normal frequencies and the forms of the normal vibrations is then the primary problem in correlating the structure and internal forces of the molecules with the observed vibrational spectrum.

The problem is rather complex. A nonlinear molecule of n atoms has $3n$ degrees of freedom, which are distributed as 3 rotational, 3 translational, and $3n - 6$ vibrational motions, each with a characteristic fundamental band frequency (linear molecules have $3n - 5$ vibrational modes). The fundamental band frequency may be defined as the absorption frequency at which a normal mode of the molecule is singly excited. Obviously, for a high polymer molecule, where n is large, the number of fundamental vibrations of the molecule can become inordinately large. Not all of these modes will be infrared active, however. No absorption will take place in the infrared, unless there is a change in the dipole moment of the molecule during the normal vibration. Thus infrared absorption will occur only for unsymmetrical vibrations.

Often, there is no change in the dipole moment of the molecule during a particular normal vibration. For example, the symmetrical stretching frequency v_1 of the linear CO_2 molecule is inactive in the infrared because the dipole moment changes produced by the two $C=O$ bonds are equal and opposite and therefore cancel.

$$\overset{\leftarrow}{O}=C=\vec{O}$$

The asymmetric stretching vibration v_3 is active in the infrared, however, with a transition moment lying along the major axis of the molecule

$$\vec{O}=\overset{\leftarrow}{C}=\vec{O}$$

because the relative motion of the atoms brings the carbon atom alternately closer to one and then to the other oxygen atom. This relative motion necessarily causes a corresponding relative motion of the charge centers. Since a change in dipole moment occurs for a molecule whenever a change in position of the centers of positive and negative charge resulting from atomic motion occurs, v_3 is infrared active.

Stretching of bonds is not the only vibrational mode of deformation in a molecule. In fact, if there is a total of $3n - 6$ vibrational modes in a molecule, $n - 1$ of these are stretching modes and $2n - 5$ are due to other deformations such as bending, wagging, twisting, and rocking. The bending mode of the CO_2 molecule, for instance,

$$\overset{\uparrow}{O}=\underset{\downarrow}{\overset{\uparrow}{C}}=\overset{\uparrow}{O}$$

has its transition moment perpendicular to the major axis of symmetry of CO_2 molecule. All of these deformation modes contribute to the vibrational spectra. Depending on the nature of the vibration, they may have transition moments (direction of change in the dipole moment) parallel, perpendicular, or at some angle to the major axis of symmetry of the molecule. This characteristic of directionality can be utilized in structure determinations.

Thus, not all of the normal modes of vibration are infrared active. To further complicate the problem of structure analysis from the observed infrared spectra, not all of the absorption frequencies observed experimentally are the result of normal modes, since the harmonic-oscillator model is an oversimplification. In reality, overtone and combination bands appear due to the effect of anharmonic terms in the potential function. An already complicated theoretical solution of the observed spectrum is thus further confused because the general potential function contains more independent force constants than can be determined from the experimental data. The number of theoretical constants can be reduced, however, by making assumptions about the nature of the force field in the molecule. One such assumption is that contributions to the potential energy arise only from the stretching of chemical bonds and from the bending of angles between chemical bonds. This is known as the valence-force-field (VFF) assumption (71). Another assumption is to include in the potential function interaction terms that represent the existence of weak central forces between nonbonded atoms (72). This is known as the Urey–Bradley force field (UBFF). Finally, a central force field (CFF), which takes into account the interactions between every pair of atoms, has also been applied in normal coordinate calculations of the vibrational spectrum to reduce the number of force constants.

The determination of the normal modes and their frequencies depends on the solution of a $3n \times 3n$ determinant. Methods exist for simplifying this computational problem that require some molecular structure to be assumed as a model. From the symmetry characteristics of the assumed molecular structure, the general determinant can be resolved into more easily computed subdeterminants of lower order (by application of group theory), each of which involves only normal frequencies of a given symmetry class. Thus, by assuming the molecular structure and the nature of the interactions (force field), the complexity of the determination of the vibrational modes of the molecule can be reduced.

Any experimental method that can help to identify the modes of motion of an observed absorption band can be of great help in resolving the validity of a normal-coordinate analysis of the observed spectra. The normal-coordinate analysis always involves more force constants than observed frequencies. While, in general, one can calculate a set of force constants that are reasonable and will reproduce the observed frequencies, the assignment of a specific frequency to a particular vibration of the molecule is not always possible (67). Similarly, since the molecular structure and force field are assumed in the calculation, independent methods for corroborating assignments are necessary in order to determine which assumed molecular structure is correct. Simply matching the observed frequencies is not a sufficient solution. Some form of independent experimental check on the calculated band assignments is necessary.

One method of qualitatively checking the assignment of a vibrational frequency as characteristic of a particular mode of motion of a group of atoms is by relating the frequency assignment for that vibrational mode to the characteristic absorption frequency of a similar group of atoms in another, known molecule. It has been demonstrated that, although a normal mode involves, to some extent, all the atoms in a molecule, a large number of these vibrations are localized to a high degree of approximation in particular bands or in small groups of atoms. These *group frequencies* persist, irrespective of the type of molecule in which the bond or group of atoms occurs. Thus, there is an invariance of the bond-stretching force constants and, therefore, the frequency of a normal vibration that is mainly confined to the stretching and contraction of one bond in the molecule may be estimated from eq. 2-47, for instance, which was derived for a diatomic molecule. Stretching frequencies calculated in this manner should be regarded as only approximate, however. Angle-bending force constants, which are generally about an order of magnitude smaller than stretching constants, and interaction constants are much more sensitive to the environment. In general, vibrational frequencies of particular groups of atoms will remain nearly constant from molecule to molecule so that some theoretical vibrational mode assignments for groups of atoms in large molecules can be checked by comparing the frequencies with those of small molecules in which the vibrations of the same groups have been determined in detail.

Another helpful method for checking spectral band assignments is isotopic substitution (70). The most useful form of isotopic substitution in a molecule is that of deuterium for hydrogen. In this case, the force constants do not change and only the masses of the hydrogen atoms are changed in magnitude. For this reason, the observed frequency of the hydrogen modes in the molecule will be reduced (see, for example, eq. 2-47) and the method, therefore, serves to identify hydrogen modes in the spectra. Similarly, bands of complex origin (where there is interaction between hydrogen modes and other modes) will usually become evident, since the interaction will generally disappear on deuteration. Obviously, if different kinds of hydrogen atom groupings are present in the molecule and they can be selectively deuterated, specific information about the assignments of these groups could be ascertained.

It was mentioned earlier that no absorption will take place in the infrared region unless there is a change in the dipole moment of the molecule during the normal vibration and, therefore, absorption will only occur for unsymmetrical vibrations. Thus, each of the infrared-active vibrational modes, be they stretching, bending, twisting, rocking, or a combination of these, will have a transition moment parallel, perpendicular, or at some angle α_v to the major axis of symmetry of the molecule (Fig. 2-22). The transition moment angle expected for each observed frequency can be calculated theoretically on the basis of the assumed molecular structure and the normal-coordinate vibrational mode assignment. If the vibrational mode assignment is correct, the theoretically calculated transition moment angle for a given observed frequency and the experimentally determined value at that frequency should be the same. The measured transition moment angle α_v is thus a desirable experimental parameter.

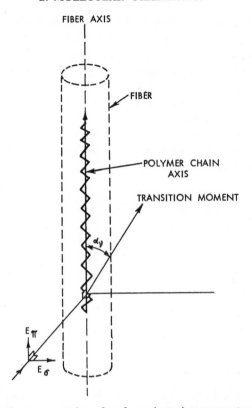

Fig. 2-22 Schematic representation of perfect orientation parameters.

The transition moment is a vector quantity and thus has both magnitude and direction. The intensity of the infrared absorption band depends upon the angle the electric vector in the incident radiation makes with the transition moment. In particular, the intensity is proportional to the square of the scalar product of the transition moment and the electric field vectors. The absorption coefficient k for a particular direction of the incident electric field vector E_0 may be written

$$k = P^2 \sum_i \cos^2 \Psi_i \qquad (2\text{-}49)$$

where Ψ_i is the angle between the transition moment direction of the ith absorbing center and the direction of the electric vector, P is the magnitude of a vector, $\mathbf{P}$ proportional to the transition moment, and the summation extends throughout unit volume of the specimen. Equation 2-49 forms the basis for the utilization of polarized infrared radiation as a powerful tool in the study of the spectra and structure of oriented polymers.

c. Infrared Measurement of the Transition Moment Angle.   The directional character of the transition moment leads one logically to the study of oriented systems for its experimental determination. The problem is similar to that of birefringence with an added complication. The orientation of the transition moments must be considered as well as the orientation of the molecules. The two-phase character of birefringence data can be avoided in infrared studies, however, at least in the initial investigation. When the polymer crystallizes into an ordered lattice, the increase in the symmetry of the molecules results in fewer and sharper bands when compared with those in the amorphous region, as in Fig. 2-23. Similarly, external lattice vibrations may even produce

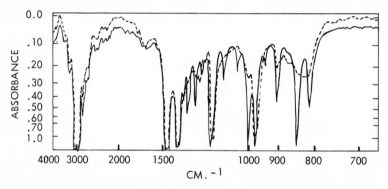

Fig. 2-23   Infrared spectra of polypropylene: — crystalline; - - - amorphous.

new bands. As a result, bands that are exclusively crystalline, bands that are amorphous, and bands that are a mixture of the contributions from both regions (mixed bands) are identifiable. Thus, by a judicious choice of absorption frequencies, the characteristics of each phase can be examined separately. Since bands characteristic of the individual phases in the film can be identified, some measure of anisotropy that can be correlated with the orientation of the phase would be desirable. The anisotropy parameter used in infrared studies is the infrared dichroic ratio.

When a polymer is uniaxially oriented, the principal axes of the refractive index ellipsoid are directed parallel and perpendicular to the fiber axis (stretch direction). When the incident electric vector is parallel to a principal axis, the radiation will traverse the specimen with its plane of polarization unaltered. Thus, only optical-density measurements made with the electric vector parallel and perpendicular to the orientation axis will be meaningful. The significant factor in interpreting the optical anisotropy of the sample in terms of molecular orientation is the dichroic ratio

$$D = \frac{k_\pi}{k_\sigma} \tag{2-50}$$

where k_π and k_σ are the principal absorption coefficients for radiation vibrating parallel and perpendicular to the direction of the orientation axis, respectively. The quantity actually measured experimentally is the dichroic ratio

$$D' = \frac{\epsilon_\pi}{\epsilon_\sigma} \tag{2-51}$$

where ϵ_π and ϵ_σ are the optical densities, $\log I_0/I$, parallel and perpendicular to the orientation axis, respectively (see Fig. 2-24), and I_0 and I are the

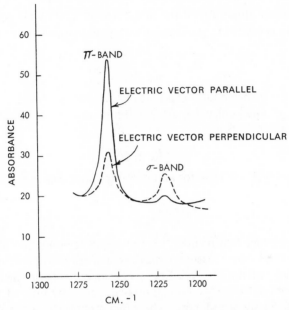

Fig. 2-24 Infrared dichroism of cast uniaxially oriented polypropylene film.

incident and transmitted intensities, respectively. The quantity D' equals D, if there are no scattering or reflection losses. Throughout the following discussion, both ϵ_π and ϵ_σ are assumed to have been corrected for non-absorptive energy losses and equal k_π and k_σ, respectively.

The relationship between the infrared dichroism D, the orientation of the molecules, and the direction of the transition moments was solved by Fraser (73). Consider a fiber where all of the polymer-chain axes are oriented parallel to the fiber axis. Optical-density measurements are made with the electric vector parallel and perpendicular to the fiber axis, respectively. For this ideal system, the incident electric vector ϵ_π (see Fig. 2-22) will be parallel to the polymer chain axis as well as to the fiber axis; hence the angle Ψ between the transition moment direction and the electric vector direction will equal

α_v. Thus, from eq. 2-49, the expression for the parallel principal absorption coefficient is

$$k_\pi = NP^2 \cos^2 \alpha_v \qquad (2\text{-}52)$$

where N is the number of absorbing centers per unit volume and α_v is the transition moment angle for the centers absorbing radiation of frequency v. From similar arguments for the perpendicular electric vector ϵ_σ the expression for the perpendicular principal absorption coefficient is

$$k_\sigma = \tfrac{1}{2} NP^2 \sin^2 \alpha_v \qquad (2\text{-}53)$$

Thus, for perfect alignment of the molecules parallel to the fiber axis, the dichroic ratio is related to the transition moment angle by the expression

$$D_0 = \frac{k_\pi}{k_\sigma} = 2 \cot^2 \alpha_v \qquad (2\text{-}54)$$

where D_0 is the dichroic ratio of an ideally oriented polymer.

When all of the molecules are randomly oriented in a film or fiber (optically isotropic), then

$$k_\pi = k_\sigma = \tfrac{1}{3} NP^2 \qquad (2\text{-}55)$$

And, thus, the dichroic ratio for an unoriented specimen is

$$D = \frac{k_\pi}{k_\sigma} = 1.0 \qquad (2\text{-}56)$$

In a real polymer film or fiber, the molecules are only partially oriented and there must be some measure of the degree of orientation of the molecules relative to the orientation axis. Fraser considered this problem from two different points of view (74). One approach, generally utilized by spectroscopists, will be discussed here. The other approach, more suitable for quantitative studies, is discussed in the next section.

An oriented polymer system can be considered as composed of a fraction f of perfectly oriented chains and a fraction $(1 - f)$ of perfectly random chains. From eqs. 2-52, 2-53, and 2-55, therefore,

$$k_\pi = NP^2 [f \cos^2 \alpha_v + \tfrac{1}{3}(1 - f)]$$
$$\text{and} \qquad (2\text{-}57)$$
$$k_\sigma = NP^2 [\tfrac{1}{2} f \sin^2 \alpha_v + \tfrac{1}{3}(1 - f)]$$

The dichroic ratio for this model, in terms of D_0 (eq. 2-54) and f, thus takes the form

$$D = \frac{1 + \tfrac{1}{3}(D_0 - 1)(1 + 2f)}{1 + \tfrac{1}{3}(D_0 - 1)(1 - f)} \qquad (2\text{-}58)$$

The following expression can then be obtained by solving eq. 2-58 in terms of the fraction of perfectly oriented chains

$$f = \frac{(D - 1)(D_0 + 2)}{(D + 2)(D_0 - 1)} \tag{2-59}$$

For any given uniaxially oriented sample, the fraction of perfectly oriented chains is a constant. The value of the constant is unknown, however. The spectroscopist is interested in determining the value of D_0 and, hence, of α_v for each absorption frequency from the measured dichroic ratios of the sample. This value of α_v then can be compared with the calculated values from the theoretical molecular models proposed. To accomplish this, the value of the unknown constant f must be determined.

One method of estimating the value of f is to use a reference band in the observed infrared spectrum. The investigator who uses this method assumes that the transition moment angle for a particular group frequency observed in the spectrum (for example, a CH_2 stretching mode) is the same as either the one calculated from the molecular structure of the polymer (75) or the one determined for a similar group vibration in a smaller model compound. By use of this assumed value of α_v and the measured value of D, the value of the constant f in eq. 2-59 is calculated. This calculated value of f then is used as a constant in eq. 2-59 for the calculation of α_v from the measured D values at the other observed absorption frequencies. A reference-band calculation of this type must be treated with some caution, since in it either the transition moment angles are assumed to be invariant from compound to compound, which assumption is questionable, or the possible effects of lattice interactions on the transition moment are ignored.

Another method for estimating f, developed by Fraser (76), permits determination of a minimum value for the fraction of perfectly oriented molecules f_m. To use this method, D_0 in eq. 59 is allowed to tend to infinity for $D > 1$ or zero for $D < 1$, giving

$$f_m = \frac{D - 1}{D + 2} \qquad \text{when } D > 1$$

$$f_m = \frac{2(1 - D)}{D + 2} \qquad \text{when } D < 1 \tag{2-60}$$

In practice, the most dichroic parallel and perpendicular bands in the observed spectrum are used to calculate the values of f_m. The greatest calculated value of f_m is then chosen for use in estimating the transition moment angle for other bands in the spectrum. By substituting the calculated f_m and the measured dichroic ratio D into eq. 2-59 a value of D_0 is obtained.

The estimated value of α_v (from eq. 2-54) is then limited to the following range:

$$\cot^{-1}\left[\frac{f_m(D + 2) + 2(D - 1)}{2f_m(D + 2) - 2(D - 1)}\right]^{1/2} \leq \alpha \leq \cot^{-1}\left(\frac{D}{2}\right)^{1/2} \quad \text{when } D > 1$$

and   (2-61)

$$\cot^{-1}\left[\frac{f_m(D + 2) - 2(1 - D)}{2f_m(D + 2) + 2(1 - D)}\right]^{1/2} \geq \alpha \geq \cot^{-1}\left(\frac{D}{2}\right)^{1/2} \quad \text{when } D < 1$$

This method for the determination of f has the advantage that neither the correctness of the molecular model nor the invariance of the transition moment angle from compound to compound must be assumed. The utility of the method in molecular conformation studies has been clearly demonstrated by Tsuboi in his study of the molecular conformation of the α form of poly-γ-benzyl-L-glutamate (77). The method has the disadvantage, however, of determining only an estimated value of f (i.e., the minimum value f_m).

The primary difficulty in the experimental determination of α_v by the infrared technique is that operationally there is one equation with two unknowns (eq. 2-59). What is needed to solve this dilemma is an independent, quantitative, experimental method for determining the unknown constant f.

d. X-Ray Diffraction and Determination of the Transition Moment Angle. In a uniaxially oriented polymer, the molecules are not ideally oriented and it is necessary to have some measure of the degree of orientation of the molecules relative to the orientation axis. Fraser (74) considered this problem from two different points of view. One of these, the one discussed above in Section B.5.c, is in general use. The other approach, discussed here, has generally been ignored. Samuels (25) has demonstrated that this neglect is unfortunate, as it is this second approach that offers a route to more quantitative structure data.

Instead of considering the oriented sample as containing a certain fraction f of perfectly oriented molecules and a fraction $(1 - f)$ of randomly oriented molecules, all of the molecules can be considered to be oriented at some average angle θ, relative to the orientation direction, as in Fig. 2-25. The dichroic ratio for this model, in terms of D_0 and the average orientation angle θ, is given by the expression

$$D = \frac{1 + (D_0 + 1)\overline{\cos^2\theta}}{1 + \frac{1}{2}(D_0 - 1)\overline{\sin^2\theta}}$$ (2-62)

which is analogous to eq. 2-58. The expression for θ is then

$$\overline{\sin^2\theta} = \frac{2(D_0 - D)}{(D_0 - 1)(D + 2)}$$ (2-63)

A further analysis of these expressions, to find the relation between the fraction of oriented chains f of the first model and the average angle of orientation θ of this model, leads to the expression

$$\frac{3\overline{\cos^2\theta} - 1}{2} = \frac{(D - 1)(D_0 + 2)}{(D + 2)(D_0 - 1)} = f \tag{2-64}$$

Thus, Fraser's fraction f is the same as Hermans' orientation function (eq. 2-9) when it is evaluated in terms of the average orientation of the polymer molecules. Samuels has shown that this deduction is very useful (25). Sections B.2–B.4 discussed various methods for determining quantitatively the crystalline and amorphous orientation functions of a uniaxially oriented, polycrystalline polymer film or fiber. Samuels recognized that any of these methods, coupled with the infrared dichroic measurements of absorption

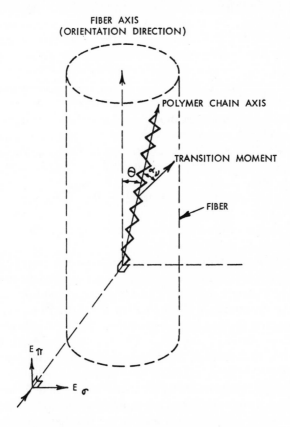

Fig. 2-25 Schematic representation of uniaxial orientation parameters.

bands from the appropriate phase, will yield quantitative values for the transition moment angles. Thus Samuels' method of experimentally determining the orientation function by an independent method, and combining this experimentally determined f with the experimentally determined dichroic ratio (D), transforms eq. 2-64 into one equation with only one experimental unknown (D_0). This equation can be explicitly solved with no approximations or assumptions required.

For example, the infrared absorption bands from the crystalline region of a uniaxially oriented polymer can be measured and the dichroic ratios determined. The orientation function for the crystalline region of the polymer film f_c can be determined quantitatively from azimuthal, wide-angle x-ray diffraction measurements of the appropriate reflecting planes of the crystal and application of Wilchinsky's method (Section B.2) for its calculation. Substitution of the measured f_c and D values into eq. 2-64 yields directly the value of D_0 and, hence, that of the transition moment angle for that absorption frequency. No assumptions are necessary in this calculation as to the correctness of a molecular model, the invariance of the transition moment angle from compound to compound, or of any approximate character for f.

Samuels also showed that an even higher degree of reliability in the experimentally determined transition moment angle can be achieved if samples with different degrees of molecular orientation are examined by both the infrared dichroic and x-ray diffraction methods (25). Equation 2-64 predicts that a plot of f_c, the crystal orientation function determined from x-ray diffraction measurements, versus $(D - 1)/(D + 2)$, where D is determined from infrared measurements, will be linear with a zero intercept. A least-squares evaluation of the data from this line should yield a value of the slope $(D_0 + 2)/(D_0 - 1)$ corrected for random errors and sample differences. A quantitative value of the transition moment angle can then be calculated from the resultant D_0 value by use of eq. 2-54.

Some results of this approach with data from the Series C uniaxially oriented, isotactic polypropylene films and from polyethylene films will now be presented to illustrate the general utility of the method. Figure 2-26 and Table 2-4 give experimental infrared data from the same cast, isotactic polypropylene films for which the x-ray diffraction, birefringence, and sonic modulus data are listed in Table 2-3. The 1220-cm^{-1} σ band is generally agreed to be due to absorption by the crystalline phase only (see Fig. 2-23), although there is no general agreement as to the assignment of the vibrational modes of the chain characterized by this band, as shown in Table 2-5. There is agreement in the vibrational-mode assignments only to the extent that the various authors (69,70,78–81) characterize the band as represented by the motion of one, two, or a mixture of three of the following bonds: CH_2, CH,

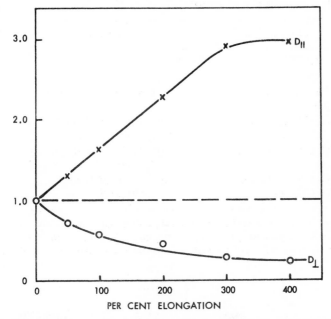

Fig. 2-26 Relation between the elongation of Series C isotactic polypropylene films, $D_\parallel$ and $D_\perp$: (O) 1220-cm^{-1} band; (×) 1256-cm^{-1} band (25).

C–CH$_3$ or an equatorial C–C. Isotactic polypropylene has a helical conformation in the crystalline phase (16) and each of these bonds makes an angle of about 72° with the helical axis as shown in Fig. 2-27.

In Fig. 2-28, the crystal orientation function f_c (from x-ray diffraction, Table 2-3) is plotted against the infrared 1220-cm^{-1} σ-band dichroic ratio according to eq. 2-64. The experimental points fit a calculated line drawn for $\alpha_v = 72°$ quite reasonably. This indicates that the transition moment is

Table 2-4. Infrared Data for Series C Isotactic Polypropylene Films (25)

Cast Sample	Elongation, %	$D_\parallel$, 1256 cm^{-1}	$D_\perp$, 1220 cm^{-1}
1	0	1.00	1.02
2	50	1.30	1.39
3	100	1.63	1.77
4	200	2.28	2.19
5	300	2.92	3.58
6	400	2.96	4.21

Table 2-5. Various Assignments for the 1220-cm^{-1} and 1256-cm^{-1} Infrared Absorption Bands of Isotactic Polypropylene[a]

$\sigma(1220\ cm^{-1})\ E$	$\pi(1256\ cm^{-1})\ A$	Ref.
$\gamma_w(C-H)$	$\gamma_w(C-H)$	69
$\gamma_w(C-H)$	$\gamma_w(C-H)$	78
$\gamma_r(CH)$ mixed with $\gamma_r(CH_2)$, $\gamma_r(CH_3)$	$\gamma_r(CH)$ mixed with $\gamma_r(CH_2)$, $\gamma_r(CH_3)$	70
$\gamma_t(CH_2)(35)$, $\delta_{ax}(CH)(15)$, $v_{eq}(C-C)(15)$	$\gamma_t(CH_2)(35)$, $\delta_{ax}(CH)(15)$, $\delta_{eq}(CH)(15)$	79
$\gamma_t(CH_2)(25)$, $\delta(CH)(15)$, $v_{eq}(C-C)(20)$	$\gamma_t(CH_2)(35)$, $\delta_{ax}(CH)(20)$, $\delta_{eq}(CH)(10)$	80
$\gamma_t(CH_2)(26)$, $\delta(CH)(20)$, $v(C-CH_3)(19)$	$\delta(CH)(37)$, $\gamma_t(CH_2)(19)$, $\gamma_r(CH_3)(15)$	81

[a] Symbols: σ, perpendicular band; π, parallel band; E, phase difference $= \pm 2\pi/3$; A, phase difference $= 0$; γ_w, wagging; γ_r, rocking; γ_t, twisting; δ, bending; v, stretching; ax, axial; eq, equatorial; numbers in parentheses indicate approximate potential energy distribution, %.

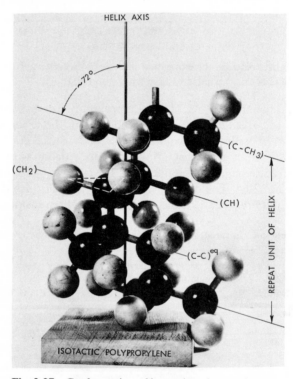

Fig. 2-27 Conformation of isotactic polypropylene.

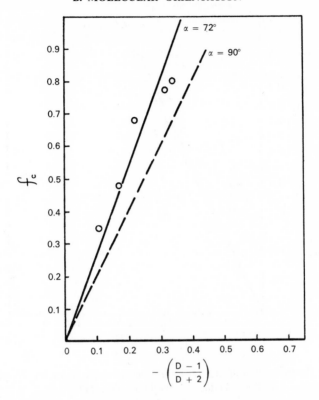

Fig. 2-28 Determination of the transition moment angle α_v for the 1220-cm^{-1} band in isotactic polypropylene (25).

directed along one of the bonds mentioned above. The dashed line in Fig. 2-28 has been calculated for $\alpha_v = 90°$ (i.e., where the transition moment of the σ band is perpendicular to the helical axis of the chain). An assignment of $\alpha_v = 90°$ for the 1220-cm^{-1} band would not be reasonable in light of the experimental data.

A further example of the general utility of this method is illustrated in Fig. 2-29, which represents the data of Norris and Stein (52) for low-density polyethylene replotted by Samuels (25) to satisfy eq. 2-64. Polyethylene has a planar zigzag conformation in the crystal lattice. The 730-cm^{-1} σ band absorbs only in the crystalline region and has been assigned as a CH$_2$ rocking mode. The α_v for this motion would be expected to be 90° to the backbone axis. From Fig. 2-29, $\alpha_v \cong 83°$ seems to be more reasonable. The observed difference in α_v between theory and experiment may be due to experimental difficulties in separating the 730–720-cm^{-1} doublet into its components, orientation effects of the amorphous region in the 720-cm^{-1}

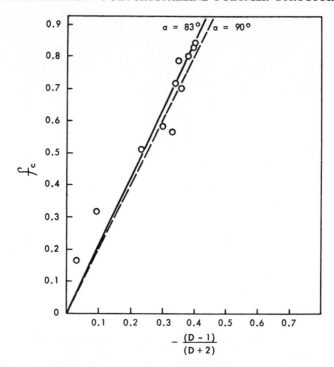

Fig. 2-29 Determination of the transition moment angle α_v for the 730-cm^{-1} band in low-density polyethylene [Data: Norris and Stein (52)] (25).

band affecting the resolution of the 730-cm^{-1} band, or possible coupling effects from bending and stretching modes.

The above method of relating orientation functions to infrared dichroism is a general one. If absorption takes place only in the crystalline phase, the α_v values can be determined quantitatively for any crystal by using Wilchinsky's method for determining f_c. This method can thus have considerable utility in structure analysis.

Absorption, however, does not always take place in only one phase of the polymer. There are some infrared absorption bands that are characterized by absorption in both the crystalline and amorphous phases. The 1256-cm^{-1} π band of isotactic polypropylene is just such a mixed band. Since this band results from absorption in both the crystalline and amorphous regions, its infrared dichroism would be expected to correlate with some average orientation function. Such an average can be expressed [Samuels (25)] as the average orientation of each phase weighted by the amount of each phase present

$$f_{av} \equiv \beta f_c + (1 - \beta)f_{am} \tag{2-4}$$

where β, the fraction of crystals, is determined from density, f_c is derived from x-ray diffraction, and f_{am} is determined from sonic modulus. Figure 2-30 is a plot of f_{av} against $(D - 1)/(D + 2)$ for the 1256-cm^{-1} π band of the oriented, isotactic polypropylene samples. A good linear plot with a zero intercept is obtained as predicted.

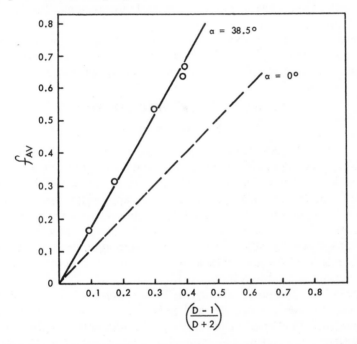

Fig. 2-30 Relation between the average orientation function f_{av} and the infrared dichroism of the 1256-cm^{-1} band in isotactic polypropylene (25).

If, in the spectra of the polymer film, absorption bands are present that are characteristic of the amorphous region only, then the infrared dichroism of these bands should correlate linearly with the amorphous orientation function according to eq. 2-64. Structural interpretation of the absorption spectra of amorphous polymers is quite complex (82) and this technique would be very helpful in characterizing assignments.

The above discussion has emphasized how quantitative structural information about polymer chains can be obtained by means of x-ray diffraction, birefringence, and sonic modulus measurements along with infrared dichroism studies on the uniaxially oriented polymer films. A primary purpose of this Chapter is to elucidate methods for the determination of the orientation characteristics of deformed polymers. Obviously, once the value of the transition moment angle for a given absorption frequency has been quantitatively determined by the above method, eq. 2-64 may be used to determine the orientation function from infrared dichroism measurements only. Thus, infrared dichroism may be used to obtain quantitative values for f_c and f_{am}, provided absorption bands characteristic of each individual phase can be obtained. If, as has been observed with isotactic polypropylene, only bands uniquely characteristic of the crystalline region are available, the infrared dichroic measurement can be combined with sonic modulus or birefringence measurements to determine f_c and f_{am} quantitatively. Infrared dichroism is thus a powerful tool for quantitatively characterizing the molecular morphology of uniaxially deformed polymer systems.

C. SPHERULITE DEFORMATION

1. Introduction

In the previous section, the polycrystalline polymer was considered as a two-phase system composed of crystallites and noncrystalline regions whose individual orientations could be followed by a number of independent physical methods. In many polymers, under appropriate processing conditions, the crystallites and noncrystalline regions will arrange themselves in an ordered superstructure called a spherulite. The polymer film will then be composed of many spherulites and any deformation process will involve the deformation of the spherulites as well as the orientation of the crystalline and noncrystalline substructure.

Polyethylene, polypropylene, nylon, Teflon, polyethylene terephthalate, polyhexamethylene sebacamide, and poly-γ-methyl-L-glutamate are some of the polymers that form one or more types of spherulites. A spherulite may be described as a spherically symmetrical aggregate of crystalline and amorphous (noncrystalline) polymer. The crystallites are arranged in large part along radial fibrils, which have a common center at the site of primary nucleation. Figure 2-31 is a phase-contrast photograph of an isotactic polypropylene spherulite. Notice the radial fibrils all emanating from the site of primary nucleation. The crystallites are aligned mainly along these radial fibrils, although some secondary crystallization can occur in the interradial region. The noncrystalline polymer can also be inter- or intraradial but is found predominantly in the interradial regions.

The intraradial crystallites generally have a preferred orientation with respect to the radial direction in the spherulite. This orientation has been determined by micro x-ray diffraction (26,83,84,85) and electron diffraction studies (86,87) of selected regions of spherulites. The polymer-chain axis in the crystallites has generally been found to be oriented perpendicular (or at some angle toward the perpendicular) to the radial direction in the spherulite and hence is tangential with respect to the spherulite. This observation has led to recent speculation that the crystallites arranged along the radii of the spherulite may be in the form of a folded-chain lamella with the chain axis oriented perpendicular to the spherulite radius.

Depending on the processing conditions, the same polymer may form spherulites having positive, negative, or mixed birefringence. The birefringence of the spherulite Δ_s is defined as $(n_r - n_t)$, where n_r is the refractive index parallel to the radial fibril and n_t is the refractive index perpendicular (tangential) to the radial fibril. If the noncrystalline polymer is assumed to be isotropic (by no means true), then the birefringence of the spherulite will be directly related to the birefringence of the crystallites along the fibril. In polymers, the largest refractive index γ is along the chain axis while the refractive indices α and β are perpendicular to the chain axis ($\beta > \alpha$).

If the crystallite is biaxial (i.e., $\beta \neq \alpha \neq \gamma$), then Δ_s can be either negative

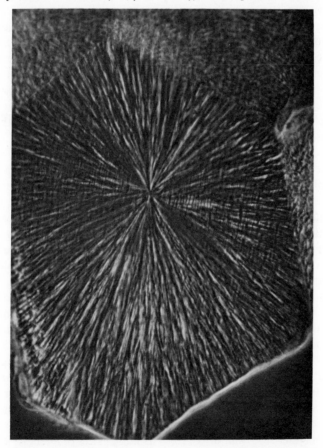

Fig. 2-31 Phase microscope photograph of an isotactic polypropylene spherulite.

or positive. For example, if the chain axis is assumed to be perpendicular to the radial axis, that is, γ is perpendicular, two possibilities can be considered: (1) If α is parallel to the radius, the spherulite birefringence Δ_s is negative because α is less than both β and γ (see above) and hence $n_t > n_r$. (2) If β is parallel to the radius, n_r can be greater or smaller than n_t. Since the chain axis is rotationally symmetrical around the radial axis, n_t will equal the average refractive index of the two (γ and α) refractive indices. Depending on the magnitude of α, then, n_t can be either greater or smaller than n_r (which is dependent only on β). If n_t is greater than n_r, the birefringence will be negative [as in (1)], but if n_t is smaller than n_r, the birefringence will be positive. Hence, the sign of the spherulite birefringence is a function of crystallite orientation.

Polyhexamethylene sebacamide provides a practical illustration. This polymer forms either negative or positive spherulites depending on the processing conditions. Micro x-ray diffraction (83) studies have shown that the chain axis of this polymer is perpendicular to the radial axis of the spherulite for both positive and negative birefringences. The a axis of the crystallite was found to be parallel to the radial direction in the positively birefringent spherulite, while the b axis of the crystallite was found to be oriented parallel to the radial direction in the negatively birefringent spherulite. The refractive index of the a axis is therefore greater than the refractive index of the b axis, and the average refractive index of the b and c axes is less than the refractive index of the a axis.

Positive and negative spherulites of the polypeptide poly-γ-methyl-L-glutamate can be obtained by careful precipitation from solution (88). The type of spherulite obtained will be a function of the time the solution is aged before precipitation occurs. Numerous types of positive, negative, mixed, or ringed spherulites can be obtained from isotactic polypropylene, depending on temperature and time of annealing (89). Similarly, many other polymers yield different spherulitic forms, depending on the conditions used in their preparation (90). Clearly, the morphology one obtains in a spherulite will be some function of the processing conditions used to make the film.

Just as spherulite morphology is a function of processing conditions, so too is the resulting size of the spherulite. The spherulites grow from primary nuclei and, hence, the size of the spherulite is governed by the number of primary nuclei present. The number of primary nuclei present is a function of the growth temperature, maximum temperature, time of melting, and the concentration of impurities present. Generally, large spherulites are formed from melts that have been heated to a high enough temperature so that any trace of previous crystallinity is lost, and from samples crystallized at low degrees of supercooling where spontaneous nucleation rates are low. The size of spherulites can range from as large as centimeters to less than a micron in diameter, depending on the processing conditions used.

The size of the spherulite dictates the method available for its measurement. Spherulites centimeters in diameter can be measured directly (with a ruler). The size of spherulites larger than a few microns in diameter can be determined with a polarizing microscope, or, within a limited size region, by means of light scattering measurements. For spherulites smaller than a few microns in diameter, the only methods of size measurement available are the electron-microscope and light-scattering techniques.

An orientation step is often included during the production of a polymer film to impart added mechanical strength to the product. If the undeformed film was spherulitic, the spherulites will be deformed during the orientation step. Depending on the speed and temperature of the orientation step, spherulite deformation may or may not be affine. An affine deformation of the spherulite is one in which a given deformation of the film will lead to an equivalent deformation of the spherulite. Takayanagi, Minami, and Nagatoshi (91) have shown that this type of deformation occurs in polypropylene when the temperature of the deformation process is at the crystal absorption temperature of the dynamic mechanical spectra (110°C) and the speed of orientation is not so rapid as to exceed the energy limits at that temperature for mobility of the crystallites. The film stretched under these conditions does not manifest necking but deforms uniformly. Films produced under these conditions had the best combination of high orientation, breaking strength, and modulus of elasticity that Takayanagi could produce. Clearly, a knowledge of spherulite size, extension ratio (i.e., amount of deformation), and the mechanism of deformation of the spherulites is desirable.

One method that can be used to obtain information about the mechanism of spherulite deformation is wide-angle x-ray diffraction. Wilchinsky (92) derived a theory relating the crystal orientation function (Section B.2) of uniaxially oriented samples to the deformation of the spherulites. Examination under the electron microscope of surfaces replicas from deformed spherulites had revealed (93) that the spherulitic shape becomes ellipsoidal with elongation. Starting with this observation as his model, Wilchinsky made the following assumptions about the deformation mechanisms in the spherulite:

1. The crystallites remain undeformed during spherulite deformation.
2. The crystallites are connected by tie molecules.
3. The chain axis of the crystallites tends to orient in the direction of the deforming strain.

He then calculated values of the crystal orientation function at various extension ratios λ for two kinds of spherulite deformation in a fiber. One (expressed as f_λ), he assumed to have uniform deformation within the spherulite; in the other ($f_{\lambda'}$), he included the enhanced orientation of the inner regions of the spherulite when it is deformed. Both were derived for simple elongation at constant volume. In both, the spherulite was assumed to increase in length by an extension ratio λ while it decreased $\lambda^{-1/2}$ in the other two orthogonal directions. Equating λ of the spherulite with λ of the deformed sample implies an affine deformation. The two models differ in the nature of the crystalline displacement within the affinely deformed spherulite.

Wilchinsky found that data from uniformly deformed polyethylene films agreed with his theoretical curve for enhanced orientation of the regions of the spherulites. The experimental values of the crystal orientation function f_c for the uniaxially deformed isotactic polypropylene films reported in Table 2-3 fall between Wilchinsky's two theoretically derived curves (Fig. 2-32). This indicates that an affine deformation of the spherulites

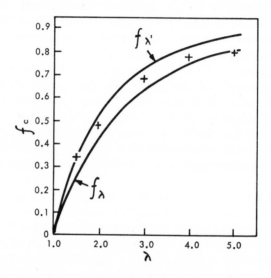

Fig. 2-32   Relation between the predicted and experimental crystal orientation functions and the extension ratio of the sample: ($-$) predicted; ($+$) experimental (25).

occurred when these isotactic polypropylene films were uniformly drawn at 110°C. The exact nature of the crystallite displacement mechanism (uniform or enhanced orientation) within the spherulites, however, is uncertain.

A much more direct method for examining the morphology of deformed spherulites is the combination of electron microscopy and electron diffraction. Generally, both of these measurements can be made with the same instrument (electron microscope), often on the same portion of the specimen. In theory, one can obtain microscopic resolution on the order of a few angstroms and, hence, small regions of deformed spherulites could be seen by means of this technique. Electron diffraction patterns then could be obtained from the same specimen region and the crystallite orientation analyzed. Theoretically, this tool should be useful in the elucidation of spherulite deformation mechanisms; practically, however, little use has been made of the technique for this purpose. The primary reasons for this lack of use are the short exposure times available and the difficulty in sample preparation.

Electron-scattering intensities are very high with respect to those of x-rays, the ratio of x-ray to electron-scattering intensities being $1:10^6$. This means that several seconds of exposure to an electron beam will yield an electron-scattering pattern equivalent in intensity to a diffraction pattern obtained from several hours of exposure to x rays. An added complication is that the polymer structure is quickly destroyed by the electron beam; therefore, little time is available experimentally to align a desired region of the sample in the microscope and obtain both a micrograph and an electron-diffraction pattern.

Similarly, the size of the specimen that can be observed must be varied according to the type of radiation used. While the linear dimension of the specimen is about 1 mm for x rays, it is only about 10^{-5}–10^{-4} mm for electrons. Polymer spherulites of general interest are of linear dimensions considerably greater than several hundred angstroms, and consequently sample preparation is a major problem in electron microscopy.

Most investigators have had to resort to either surface-replication techniques (90,93–95), which allow only observation of the surface deformation of the spherulites, or fracture-surface studies (96) to obtain specimens for observation. Ingram and Peterlin (86) have shown how this limitation could in part be overcome by preparing solution-cast polyethylene spherulites on Mylar film. The Mylar film with the attached spherulite (two dimensional) was then deformed 30%, backed with carbon, shadowed, and examined by electron-microscopy and -diffraction techniques. In this manner, the authors were able to show evidence for the drawing out of oriented fibers from interspherulitic boundaries, tilting and/or rotation of the crystallographic axes of lamella regions, lamella slip, and possible phase changes during deformation. Similarly, Selikhova, Markova, and Kargin (87) have been able to obtain electron-diffraction patterns from the radial fibrils of polypropylene spherulites which agree with micro x-ray diffraction results (26,84,85).

These techniques are limited by the fact that bulk samples of three-dimensional spherulites, which are the specimens of most practical interest, could not be examined. Bassett (97) has developed a selective dissolution technique which allows thin surface layers from bulk samples to be detached for study in the electron microscope. The resulting detachment replica diffracts electrons and, hence, the same region can be examined by electron-diffraction and -microscopy techniques. The problem with a technique of this type is that it is impossible to know whether the detachment replica under observation actually represents the structure present in the bulk sample or if the dissolution procedure produced morphological changes in the specimen.

Another approach to the problem of bulk sample preparation is that of ultramicrotoming sections for examination. Rusnock and Hanson (98) have developed a sample-embedding technique that allowed them to ultra-microtome 500-Å sections of nylon 66 spherulites from molding pellets and filaments and examine them under the electron microscope. No attempt was made to obtain electron-diffraction patterns from the samples, although there seems no reason why this should not have been possible. Attempts should certainly be made to extend this technique to other polymers. Clearly, sample preparation is a major problem in the application of electron-microscopy and -diffraction techniques to the study of spherulite deformation mechanisms.

An attractive extension of the electron-diffraction technique has been developed by Bassett and Keller (99). From detachment replicas of bulk polyethylene films, they have obtained small-angle electron-diffraction patterns, which are similar in form to small-angle x-ray diffraction patterns. Spacings as large as 1400 Å were observed from replicas of sections of drawn, linear polyethylene films which had originally contained ringed spherulites in the undrawn film. These spacings were obtained from replica regions which were observed to contain a quasiperiodic arrangement of dark lines having separations that varied between 500 and 5000 Å in different regions in the field. The small angle spacing was therefore interpreted as arising from the quasiperiodic structure. Thus, wide-angle electron diffraction, small-angle electron scattering, and visible observations of small regions of bulk samples can be obtained with the electron microscope.

The electron microscope will certainly contribute significant information on deformation mechanisms in spherulitic polymers. The technique, however, is a destructive one, that is, the sample to be examined must be destroyed during its examination. To be able to observe the deformation behavior of spherulites in bulk polymer samples under controlled conditions of tem-perature and stress without destruction of the sample is highly desirable. Small-angle light scattering (SALS) is the only system now available for quantitatively satisfying this condition. Bulk samples up to 10 mils thick and spherulites ranging from 0.1 to 50 μ in diameter have been examined non-destructively by means of this technique. By examining changes in the light-scattering patterns at small angles from the incident beam, as will be described in the following section, information can be obtained about the size and deformation of the spherulites in the sample. For this reason, the SALS technique is an important tool for elucidation of the morphology of deformed spherulitic fibers and films.

2. Small-Angle Light Scattering (SALS)*

a. Undeformed Spherulites. A spherulite is a three-dimensional, spherically symmetrical aggregate of crystalline and noncrystalline polymer. Any theory which purports to describe small-angle light-scattering (SALS) behavior of spherulites must use as its model the known three-dimensional structure of the spherulite as observed by light and electron microscopy. The amplitude method (100) for calculating the intensity envelope of scattered radiation is most useful for the theoretical small-angle light-scattering approach since it requires a model for its derivation. By choosing a model that realistically represents the known morphological characteristics of the spherulites, the observed small-angle light-scattering behavior of spherulitic film should be predictable.

Typical SALS patterns obtained from undeformed, isotactic polypropylene spherulites are shown in Fig. 2-33. These patterns were observed photographically by means of the system illustrated schematically in Fig. 2-34 (101). A continuous-wave He–Ne gas laser is used as the polarized, monochromatic ($\lambda = 6328$ Å) light source. The vertically polarized incident light from the laser, E_0 impinges on the sample and is scattered by the spherulites. For uniaxially deformed samples, the stretch direction (SD) is aligned parallel to the plane of the incident light (Z). The scattered ray E passes through the analyzer and impinges on the flat-plate photographic film. The position of the scattered reflection is defined by two angles θ and μ. The radial angle θ is defined as the angle between the unit vectors S_0 and S, which specify the direction of the incident and scattered beams, respectively. The azimuthal (tilt) angle μ is defined as the angle between the Z axis (stretch direction) and the projection of the difference $S - S_0$ between the unit vectors on the yz plane.

An analyzer is placed between the sample and the photographic film in Fig. 2-34. When the plane of the analyzer is horizontal (i.e., the plane of the analyzer and the plane of the incident light are perpendicular), an H_v SALS pattern is obtained (Fig. 2-33a). By convention, the subscript in the term H_v defines the plane of the incident polarized light, which here is vertical. If the plane of the analyzer is vertical (i.e., the plane of the incident polarized light and the plane of the analyzer are parallel) a V_v pattern is obtained (Fig. 2-33b).

Since the undeformed spherulite is spherical, the simplest model to apply to the amplitude method to describe the observed scattering behavior would be that of a uniform isotropic sphere of polarizability α_0 and radius R_0.

---

* For a more detailed discussion of small-angle light scattering from anisotropic spheres, disks, and rods see R. J. Samuels, *Small-Angle Light Scattering from Polymers*, in preparation.

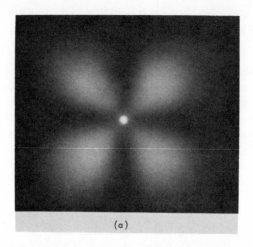

(a)

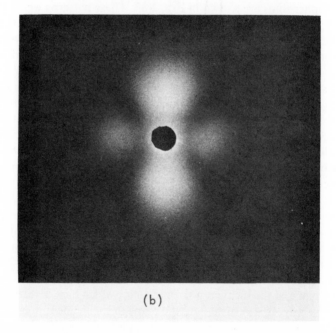

(b)

Fig. 2-33 Typical SALS patterns from unoriented isotactic polypropylene film. The polarization direction is vertical: (a) H_v; (b) V_v (101).

Starting with this model, the following expression is obtained for the scattered intensity (100):

$$I = AV_0^2\alpha_0^2\left[\left(\frac{3}{U^3}\right)(\sin U - U \cos U)\right]^2 \qquad (2\text{-}65)$$

where I is the intensity, A is a proportionality constant, V_0 is the volume of the isotropic sphere, the shape factor is

$$U = \frac{4\pi R_0}{\lambda'} \sin \frac{\theta}{2} \qquad (2\text{-}66)$$

λ' is the wavelength of light in the medium, and θ is the polar scattering angle.

The undeformed spherulite is not isotropic, however (see Section C.1); instead, it has different radial and tangential refractive indices due to the ordered arrangement of anisotropic crystallites along its radii. Thus, a more reasonable model with which to represent the undeformed spherulite is that of an anisotropic sphere in an isotropic medium.

Stein and Rhodes (102) considered the problem of the SALS patterns to be expected from a homogeneous anisotropic sphere in an isotropic medium. The correct forms (103) of the SALS equations for the intensity of the scattered light from a V_v or H_v measurement are:

$$I_{V_v} = A\rho^2 V_0^2\left(\frac{3}{U^3}\right)^2 [(\alpha_t - \alpha_s)(2 \sin U - U \cos U - \text{Si } U)$$

$$+ (\alpha_r - \alpha_s)(\text{Si } U - \sin U)$$

$$+ (\alpha_r - \alpha_t) \cos^2 (\theta/2) \cos^2 \mu (4 \sin U - U \cos U - 3 \text{Si } U)]^2 \qquad (2\text{-}67)$$

$$I_{H_v} = A\rho^2 V_0^2\left(\frac{3}{U^3}\right)^2$$

$$[(\alpha_r - \alpha_t) \cos^2 (\theta/2) \sin \mu \cos \mu (4 \sin U - U \cos U - 3 \text{Si } U)]^2 \qquad (2\text{-}68)$$

where V_0 is the volume of the anisotropic sphere, α_t and α_r are the tangential and radial polarizabilities of the sphere, α_s is the polarizability of the surroundings, θ and μ are the radial and azimuthal scattering angles, respectively (Fig. 2-34), and A is a proportionality constant. Si U is the sine integral defined by

$$\text{Si } U = \int_0^U \frac{\sin x}{x} dx \qquad (2\text{-}69)$$

and is solved as a series expansion sum for computational purposes; U has the same definition as in the isotropic sphere model (2-66),

$$U = \frac{4\pi R_0}{\lambda'} \sin \frac{\theta}{2} \qquad (2\text{-}70)$$

except that R_0 is now the radius of the anisotropic sphere. The quantity U has the same definition for both models since they both are spherical and U depends only on the shape of the model. ρ is a geometric polarization

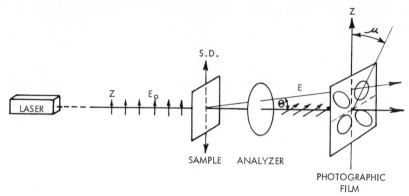

Fig. 2-34   Diagram of the photographic SALS system (101).

correction term defined as

for V_v SALS $\rho = \cos\theta(\cos^2\theta + \sin^2\theta\cos^2\mu)^{-1/2}$

for H_v SALS $\rho = \cos\theta(\cos^2\theta + \sin^2\theta\sin^2\mu)^{-1/2}$

In the limit of isotropic spheres, eqs. 2-67 and 2-68 reduce to eq. 2-65.

Figure 2-35 contains the predicted H_v and V_v SALS patterns for undeformed, isotactic polypropylene spherulites (103), derived by using eqs. 2-67 and 2-68 with the shape factor defined by eq. 2-70. These theoretical SALS patterns are plotted as contour lines of constant intensity. The contour line of squares is the line of highest intensity. The contour lines then decrease in intensity going out from the highest intensity (squares) to an intermediate intensity (crosses) to the lowest intensity (triangles). The spherulite birefringence $\Delta_s = (n_r - n_t)$ is taken to be -0.003 for calculation purposes (26,89) and, hence, an $(\alpha_r - \alpha_t)$ of -0.003 was used in eqs. 2-67 and 2-68. In order to derive these equations, α_t, α_r, and α_s had to be assumed to be sufficiently close in magnitude that the incident light wave is not affected appreciably on passing through the sphere boundary, while the large dependency of the observed V_v SALS intensity on the azimuthal angle μ indicated that the coefficient $(\alpha_r - \alpha_t)$ in the third term of eq. 2-67 must be as large as or larger than the coefficients $(\alpha_t - \alpha_s)$ and $(\alpha_r - \alpha_s)$ of the first and second terms (102,103). Thus, the difference between the polarizability of the spherulite and its surroundings must be equal to or less than the anisotropy of the spherulite. Since the surroundings of a polypropylene spherulite in a film are other polypropylene spherulites, these conditions seem reasonable. In calculating the theoretical V_v SALS pattern in Fig. 2-35, the polarizability of the surroundings α_s was assumed to be equal to the average polarizability of the spherulite. The agreement between the theoretical SALS patterns in Fig. 2-35 and the experimental SALS patterns in Fig. 2-33 demonstrates that

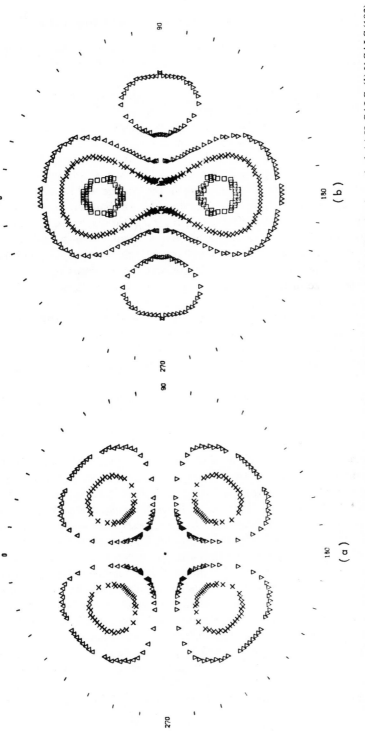

Fig. 2-35 Theoretical SALS patterns for unoriented isotactic polypropylene film. The polarization direction is vertical: (a) H_v SALS; (b) V_v SALS (103).

the model of the spherulite as an anisotropic sphere in an isotropic medium validly represents the observed SALS behavior of undeformed, isotactic polypropylene spherulites.

The V_v SALS and H_v SALS patterns each yield different information about the spherulite structure. Samuels (103) has shown that the shape of the V_v SALS pattern will depend on the refractive index of the surroundings. When the refractive index of the surroundings matches the average of the spherulite refractive indices, Fig. 2-35b will result. As the refractive index of the surroundings is changed with respect to that of the spherulite the V_v SALS pattern will change. The character of the change will be determined by the optical sign of the spherulite. The direction of change of the V_v SALS pattern with changing surrounding refractive index for a negative spherulite will be opposite that observed for a positive spherulite. Thus small-angle light scattering can be a useful tool for characterizing the optical sign of the birefringence of spherulites too small to measure in the microscope.

The average size of the undeformed spherulites can be determined from an analysis of the H_v SALS pattern. The intensity of the H_v SALS from undeformed spherulites will go through a maximum with increasing radial scattering angle $\theta/2$ independent of the azimuthal angle examined, as illustrated in a later discussion of photometric measurements of H_v SALS patterns (see Fig. 2-43, $\lambda = 1.0$). The value of U in eq. 2-70 at this $(\theta/2)_{max}$ equals 4.09. Thus, the average radius R_0 of the anisotropic spheres (undeformed spherulites) can be obtained from a determination of $(\theta/2)_{max}$ from the H_v SALS pattern and subsequent solution of eq. 2-70.

b. Deformed Spherulites

(1) Theory.   One of the unique characteristics of the SALS technique is that with it the deformation behavior of spherulites can be followed non-destructively under controlled conditions of temperature and stress. The theoretical interpretation of the SALS behavior of undeformed spherulites extended the isotropic scattering equation (eq. 2-65) specifically to include the anisotropy of polarizability. To interpret the SALS behavior of deformed spherulites, the anisotropy of the shape of the spherulite imposed by deformation must also be considered. The small-angle light-scattering theory has been extended to uniaxially deformed spherulites (101). As a result, the observed changes in the SALS patterns with deformation can be interpreted theoretically in terms of the size of the original undeformed spherulite and the subsequent deformation the spherulites have undergone.

The manner in which the H_v SALS pattern changes with spherulite deformation has been treated by Samuels in two ways; (1) qualitatively (25), and (2) quantitatively (101).

Qualitatively, the changes in the H_v SALS patterns shown in Fig. 2-36,

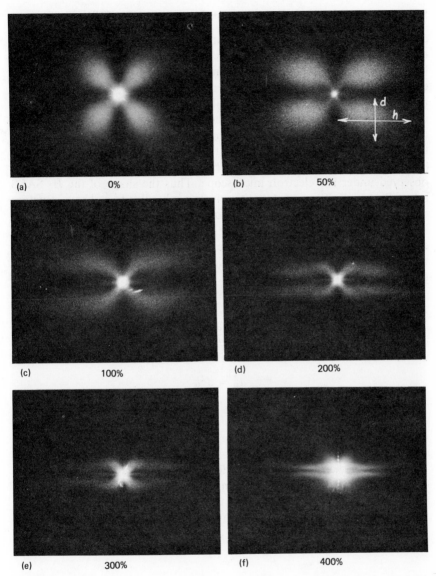

Fig. 2-36 The change in the H_v SALS pattern of Series C isotactic polypropylene film with elongation. The polarization direction and the film-stretch direction are vertical.

can be explained as follows. When a sample is stretched uniaxially, the shape of the H_v SALS lobe changes from circular to ellipsoidal. The change in the lobe scattering envelope takes the form of an extension of the lobe perpendicular to the polarization direction and a shrinkage of the lobe parallel to the polarization direction. This is illustrated in Fig. 2-36b, in which the extension is along the major axis of the ellipse h and the shrinkage is along the minor axis of the ellipse d. As the lobe of the H_v SALS pattern goes through a deformation from a circular to an elliptical to an extended elliptical (rodlike) form, there is a very similar transition in the shape of the spherulite observed under the electron microscope. Thus the shape of the H_v SALS lobe reflects the continuous transition from a spherulite to a fibrous structure occurring in the sample when it is deformed.

If any change in the shape of the spherulite is assumed to lead to an equivalent change in the shape of the H_v SALS lobe, an equation relating the shape of the H_v SALS lobe to the average extension ratio of the spherulites producing that lobe λ_s can be derived.

The unstretched spherulite has a volume $V_U = (\pi/6)d_0^3$, where d_0 is the diameter of the sphere.

The deformed spherulite is a prolate spheroid, $V_D = (\pi/6)d^2h$, where h is the length of the major axis and d the length of the minor axis. In the unstretched state, $V_U = V_D$, and therefore:

$$d_0^3 = d^2h$$

Now during extension $d_0 \rightarrow (\lambda_s d_0)$ in the stretch direction and, therefore, $\lambda_s d_0 = h$; here λ_s is the extension ratio of the spherulite; also $d_0 \rightarrow \lambda_s^{-1/2}d_0$ in the width (minor axis) direction for simple extension at constant volume, and obviously then $d = \lambda_s^{-1/2}d_0$.

To take our pictures, the film holder was placed at an arbitrary position for each picture, and hence the sample-to-film distances varied. For this reason a ratio of (h/d) was required to cancel out effects of sample-to-film variations. On substitution of the above expressions for h and d one obtains:

$$\frac{h}{d} = \frac{\lambda_s d_0}{d_0 \lambda_s^{-1/2}} = \lambda_s^{3/2}$$

or the extension ratio of the spherulite:

$$\lambda_s = (h/d)^{2/3} \qquad (2\text{-}71)$$

The extension ratio λ_s for the spherulite can be equated to the extension ratio of the uniaxially deformed film λ if an affine deformation of the spherulites has occurred. Figure 2-37 is a plot of λ_s, obtained from the SALS patterns in Fig. 2-36, against the sample extension ratio λ. The experimental points fit the predicted line for an affine deformation of the spherulites. Thus,

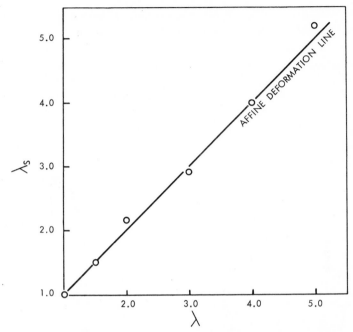

Fig. 2-37 Relation between λ_s from SALS and the sample extension ratio λ.

both a qualitative evaluation of SALS measurements and the wide-angle x-ray diffraction analysis have shown that the spherulites in the Series C film deform affinely with the sample when it is elongated at 110°C.

A quantitative evaluation of the effect of spherulite deformation on the SALS patterns has also been made by Samuels (101). Here again it is recognized that when a spherulite undergoes a uniform uniaxial deformation, it changes its shape from a sphere to a prolate spheroid. The axial ratio of the resulting spheroid is a function of the extension of the spherulite. Thus, the shape of the model used to represent the small-angle light scattering from a deformed spherulite must be changed from a sphere to a prolate spheroid.

A change in the model from a sphere to a spheroid results in a change in the definition of the shape factor U in eqs. 2-65 to 2-68. The model used is an ellipsoid of revolution with semiaxes R and vR. The shape factor for this model is defined by the expression (104–107)

$$U = \frac{4\pi R}{\lambda'} \sin\left(\frac{\theta}{2}\right)(\sin^2 \Psi + v^2 \cos^2 \Psi)^{1/2} \qquad (2\text{-}72)$$

where λ' is the wavelength of light in the medium, θ is the radial angle defined previously, and Ψ is the angle between the Z axis (stretch direction) and the vector $\mathbf{S}$ (the difference $\mathbf{S} - \mathbf{S}_0$ between the unit vectors). Since the relation between the angle Ψ and the azimuthal angle μ (107) is $\cos \Psi = \cos (\theta/2) \cos \mu$, the expression for U can be written in the form

$$U = \frac{4\pi R}{\lambda'} \sin\left(\frac{\theta}{2}\right)\left[1 + (v^2 - 1) \cos^2\left(\frac{\theta}{2}\right) \cos^2 \mu\right]^{1/2} \quad (2\text{-}73)$$

For eq. 2-73 to be useful, R and v must be defined in terms of the extension ratio λ_s of a uniaxially deformed spherulite at constant volume.

When a spherulite of radius R_0 undergoes a simple extension at constant volume, the radius is extended $\lambda_s R_0$ in the direction of extension while it contracts $\lambda_s^{-1/2} R_0$ in the two perpendicular directions (Fig. 2-38). The

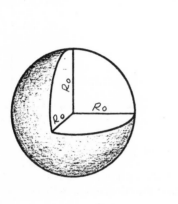

Fig. 2-38 Schematic representation of spherulite deformation.

resulting deformed spherulite is a prolate spheroid with semimajor axis $\lambda_s R_0$ and semiminor axes $\lambda_s^{-1/2} R_0$. The model used to derive eq. 2-73 was an ellipsoid of revolution with semimajor axis vR and semiminor axes R. By identifying the dimensions of the prolate spheroid (spherulite) with those of the model used to derive eq. 2-73, the following expressions are obtained:

$$R \equiv \lambda_s^{-1/2} R_0$$
$$vR \equiv \lambda_s R_0 \quad (2\text{-}74)$$
$$v \equiv \lambda_s^{3/2}$$

Substituting these definitions for R and v into eq. 2-73, the following final form for the shape factor is obtained:

$$U = \frac{4\pi R_0 \lambda_s^{-1/2}}{\lambda'} \sin\left(\frac{\theta}{2}\right)\left[1 + (\lambda_s^3 - 1) \cos^2\left(\frac{\theta}{2}\right) \cos^2 \mu\right]^{1/2} \quad (2\text{-}75)$$

The deformed spherulite is an anisotropic spheroid and eqs. 2-67 and 2-68 represent the behavior of an anisotropic spheroid in an isotropic medium when U is defined by the general shape factor given in eq. 2-75. Thus, eqs. 2-67 and 2-68, U being defined by eq. 2-75, are general equations describing the scattering from anisotropic spheroids in an isotropic medium. The models of an anisotropic sphere in an isotropic medium, and an isotropic sphere, represent limiting cases of the general equations.

(2) Photographic SALS Measurements. Flat-plate photographs of the H_v SALS patterns from the same Series C isotactic polypropylene films from which all of the data in Tables 2-2 to 2-4 were obtained are shown in Figs. 2-39 and 2-40. The photographs were obtained with the experimental arrangement shown schematically in Fig. 2-34. Each of the H_v SALS patterns for the affinely deformed films has a matching theoretical pattern alongside. The general form of eq. 2-68 (i.e., U defined by eq. 2-75) was used to calculate the theoretical H_v SALS patterns to be expected for a spherulite of a given initial radius R_0 as it is extended uniformly to the different extension ratios λ_s. The theoretical H_v SALS patterns were calculated for a spherulite having an initial radius $R_0 = 0.66$ μ and a value of $(\alpha_r - \alpha_t) = -0.003$. The agreement between the experimentally determined and theoretically calculated patterns is evident, particularly the following features:

1. The change in shape of the clover leaf patterns with elongation of the experimental and theoretical figures is the same.

2. The theoretical lobes are drawn with contour lines of constant intensity. The center contour line of the lobe is the one of highest intensity, while the outermost line of the lobe is that of lowest intensity. The corresponding contour lines from pattern to pattern are of identical intensities. As the elongation increases, the high-intensity regions move to lower radial angles while the lower-intensity contour line stays elongated to a high radial angle. The experimental flat-plate photographs of the H_v SALS pattern show this theoretically predicted intensity behavior. There is an increase in the concentration of intensity at low radial angles of the lobe and a decrease in the intensity of the extended wings of the lobe at higher radial angles with increasing elongation of the samples.

Similarly, by using the general form of eq. 2-67 (with U defined by eq. 2-75) one may calculate the theoretical V_v SALS patterns to be expected from a spherulite of a given radius R_0 as it is extended uniformly to different extension ratios λ_s. The calculated V_v SALS patterns for a spherulite having an initial radius $R_0 = 0.66$ μ, $\alpha_r - \alpha_t = -0.003$, $\alpha_t - \alpha_s = 0.000$, and $\alpha_r - \alpha_s = -0.003$ are illustrated for several elongations in Fig. 2-41.

Here, again, the theoretical lobes are drawn with contour lines of constant intensity. The outermost continuous line in each lobe is the one of weakest

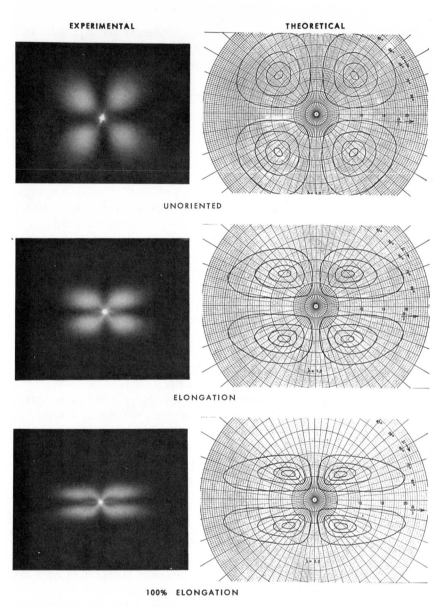

EXPERIMENTAL   THEORETICAL

UNORIENTED

ELONGATION

100% ELONGATION

Fig. 2-39 Theoretically predicted and experimentally determined changes in the H_v SALS pattern of Series C isotactic polypropylene film with elongation. The polarization direction and the film-stretch direction are vertical (101).

EXPERIMENTAL   THEORETICAL

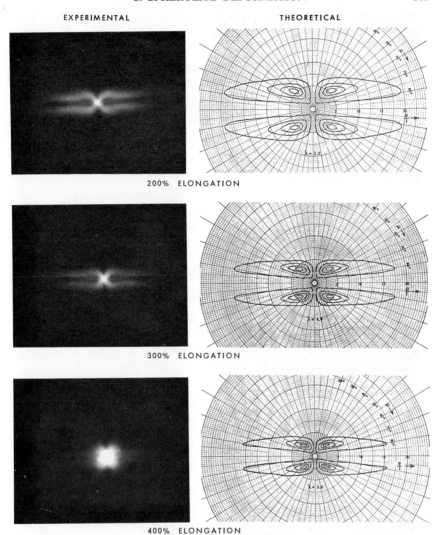

Fig. 2-40 Theoretically predicted and experimentally determined changes in the H_v SALS pattern of Series C isotactic polypropylene film with elongation. The polarization direction and the film-stretch direction are vertical (101).

intensity, the dashed line is next strongest, and the inner continuous line is of highest intensity. This highest-intensity lobe is found only near the center of the pattern and is not found in the vertical lobes. The corresponding contour lines from pattern to pattern are of identical intensities. At the side of each calculated pattern is an experimental V_v SALS pattern obtained from the

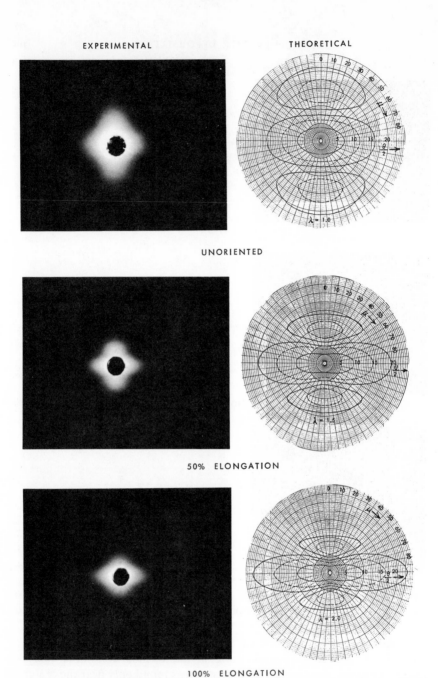

EXPERIMENTAL   THEORETICAL

UNORIENTED

50% ELONGATION

100% ELONGATION

Fig. 2-41 Theoretically predicted and experimentally determined changes in the V_v SALS pattern of Series C isotactic polypropylene film with elongation. The polarization direction and the film-stretch direction are vertical (101).

same isotactic polypropylene cast films ($R_0 = 0.5$–$0.7\ \mu$) used to obtain the H_v SALS patterns.

The black spot in the center of each pattern is caused by a beam stop, which was found necessary to prevent overexposure due to the high intensity at the center of the pattern. The agreement between the theoretically predicted and experimentally observed V_v SALS patterns is striking. The vertical lobes of the pattern obtained from the undeformed spherulites ($\lambda = 1.0$) extends to higher radial ($\theta/2$) angles than does the horizontal lobe. On deformation of the spherulite by as little as 50% elongation, the situation is reversed and the horizontal lobe now extends to a higher radial angle than that of the vertical lobes. Continued extension of the spherulites to 100% elongation serves to further increase the disparity in radial angle between the horizontal and vertical lobes.

Thus, by using a model which is consistent with the known structure of the spherulites (i.e., an anisotropic spheroid in an isotropic medium), the photographically observed SALS behavior of uniaxially deformed spherulites can be described theoretically.

(3) *Photometric SALS Measurements.* A theoretical explanation of the observed SALS patterns from deformed films gives the experimenter considerable insight into the effect of processing conditions on the behavior of the spherulites. A quantitative numerical measurement of the deformation the spherulite has undergone would be even more useful to him. With the equations for the anisotropic spheroid, a method can be developed for determining both the initial radius R_0 and the extension ratio λ_s from an analysis of the small-angle light-scattering pattern from deformed spherulites.

Due to the dependency of the terms $(3/U^3)(4 \sin U - U \cos U - 3\,\mathrm{Si}\,U)$ in eq. 2-68 on U, a maximum intensity will always be observed at a value of $U = 4.09$, provided $[\cos^2 (\theta/2)]^2$ in eq. 2-68 has a value of unity. Thus, if the radial intensity distribution at a fixed azimuthal angle μ_1 is measured, a maximum will be found in the plot of $I_{H_v}/[\cos^2 (\theta/2)]^2$ versus $\theta/2$. The value of U at this ($\theta_{\max,1}/2$) at the fixed μ_1 will be 4.09. Under these conditions, eq. 2-75 can be written in the form:

$$\frac{4\pi R_0 \lambda_s^{-1/2}}{\lambda'} \sin\left(\frac{\theta_{\max,1}}{2}\right)\left[1 + (\lambda_s^3 - 1)\cos^2\left(\frac{\theta_{\max,1}}{2}\right)\cos^2 \mu_1\right]^{1/2} = 4.09\,.$$

$$(2\text{-}76)$$

If this is repeated at a second azimuthal angle μ_2 to determine ($\theta_{\max,2}/2$), eq. 2-75 may be written as

$$\frac{4\pi R_0 \lambda_s^{-1/2}}{\lambda'} \sin\left(\frac{\theta_{\max,2}}{2}\right)\left[1 + (\lambda_s^3 - 1)\cos^2\left(\frac{\theta_{\max,2}}{2}\right)\cos^2 \mu_2\right]^{1/2} = 4.09$$

$$(2\text{-}77)$$

By solving eqs. 2-76 and 2-77 simultaneously, the following expression for λ_s is obtained:

$$\lambda_s = \left[1 + \left(\frac{4[\sin^2 (\theta_{max,1}/2) - \sin^2 (\theta_{max,2}/2)]}{\sin^2 \theta_{max,2} \cos^2 \mu_2 - \sin^2 \theta_{max,1} \cos^2 \mu_1} \right) \right]^{1/2} \qquad (2\text{-}78)$$

The original undeformed radius of the spherulite R_0 may then be calculated by substitution of the value of the extension ratio of the spherulite λ_s obtained from eq. 2-78 into eq. 2-76 or 2-77. Thus, in theory, both the extension ratio and the original radius of the undeformed spherulite can be determined from an analysis of the radial intensity distribution at two azimuthal angles of the H_v SALS pattern from deformed spherulites.

An experimental arrangement that has been used by the author to obtain the radial intensity distribution at different azimuthal angles is illustrated schematically in Fig. 2-42 (101). The plane of polarization of the laser beam

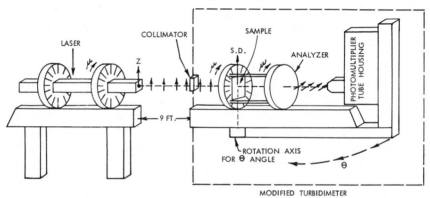

Fig. 2-42   Diagram of the photometric SALS system (101).

is fixed parallel to the stretch direction of the sample (for an oriented film) and the plane of polarization of the analyzer is perpendicular to both the plane of the laser beam and the stretch direction of the sample. The orientations of the laser beam, the sample, and the analyzer are fixed in this relationship. When the H_v SALS pattern is tilted at an angle μ all three components (the laser, sample, and analyzer) are tilted together. The photomultiplier measures I_{H_v} (the intensity of the H_v scattered radiation) as a function of the radial angle θ at each tilt angle μ. A background measurement is made for each sample at $\mu = 0°$. The background is then subtracted from the intensity I_{H_v} (measured) obtained at each tilt angle μ to give the final I_{H_v} used for interpretation.

The above theory predicts that both the extension ratio of the spherulite λ_s and the original radius of the undeformed spherulite R_0 can be determined from an analysis of the radial intensity distribution at two azimuthal angles of the H_v SALS pattern and substitution of the results into eqs. 2-78 and 2-76 or 2-77. The radial intensity distribution at different azimuthal angles μ can be calculated theoretically from eqs. 2-68 and 2-75 for isotactic polypropylene samples with given values of R_0 and λ_s. Figure 2-43 contains the theoretically calculated radial intensity distributions for spherulites with an initial radius of the undeformed spherulite of $R_0 = 0.66\,\mu$, $\alpha_r - \alpha_t = -0.003$, and $\lambda_s = 1, 1.5$, and 3 at azimuthal angles of $\mu = 20, 35, 45$, and $55°$. For the undeformed spherulite, the extension ratio λ_s equals unity and $\theta_{max}/2$ for all azimuthal angles μ has the same value (follows eq. 2-70). The calculated $\theta_{max}/2$ for the given λ_s and R_0 values is designated by the numbered line at the top of each peak. At 50% elongation of the spherulite ($\lambda_s = 1.5$), the radial curves have a different $\theta_{max}/2$ for each azimuthal angle (follows eq. 2-75). The peak intensity is still at $\mu = 45°$ but the position of the peak has shifted to a lower $\theta_{max}/2$ than was calculated for the undeformed spherulite. As expected, the position of $\theta_{max}/2$ for $\mu = 45°$ has shifted to an even lower radial angle when the spherulite has been elongated 200% ($\lambda_s = 3.0$), and the $\theta_{max}/2$ values for the other azimuthal angles are all different from each other.

Figure 2-44 contains the experimentally determined radial intensity distribution curves obtained from isotactic polypropylene cast spherulitic films that were affinely deformed to extension ratios of $\lambda_s = 1, 1.5$, and 3.

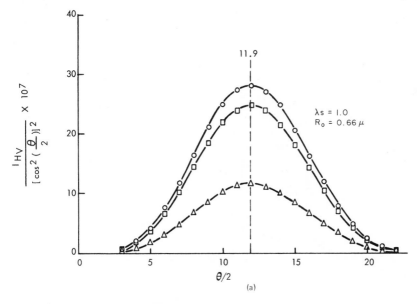

Fig. 2-43 Theoretical H_v SALS intensity distribution curves: ($\triangle$) $\mu = 20°$, $70°$; ($\square$) $\mu = 35°, 55°$; ($\bigcirc$) $\mu = 45°$ (101).

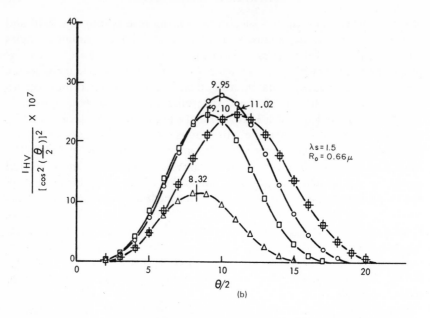

(b)

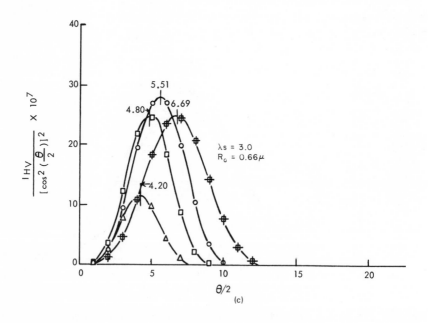

(c)

These are the same films from which the SALS photographs were obtained. The similarity between the theoretically predicted curves (Fig. 2-43) and the experimental curves is obvious.

The intensity magnitudes on the ordinate of Fig. 2-43 have not been scaled to those in Fig. 2-44 (i.e., the proportionality constant A in eq. 2-68 has not been used). The $\theta_{max}/2$ value for the undeformed sample in Fig. 2-44 is the same for all values of the azimuthal angle μ as expected. Similarly, the $\theta_{max}/2$ values for the $\lambda_s = 1.5$ and $\lambda_s = 3.0$ samples have different values for different azimuthal angles. Since the extension ratio λ_s was known for the isotactic polypropylene films examined, $\theta_{max}/2$ could be calculated for the experimentally determined intensity distribution curves at each extension ratio. The numbered line at the top of each peak is the calculated $\theta_{max}/2$ value for the given λ_s and R_0 values. The agreement between the experimental and calculated maxima is excellent. In all of these respects, the experimentally observed SALS intensity distributions from the isotactic polypropylene films agree with those predicted by theory.

Thus, a quantitative numerical determination of the deformation the spherulite has undergone during mechanical or thermal treatment of the film can be obtained from H_v SALS photometric measurements. To do so, these measurements are treated theoretically by assuming that the deformed spherulite is an anisotropic spheroid in an isotropic medium, a model

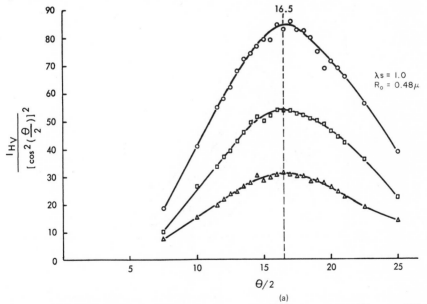

(a)

Fig. 2-44 Experimental H_v SALS intensity distribution curves: ($\triangle$) $\mu = 20°$; ($\square$) $\mu = 35°$; ($\bigcirc$) $\mu = 45°$ (101).

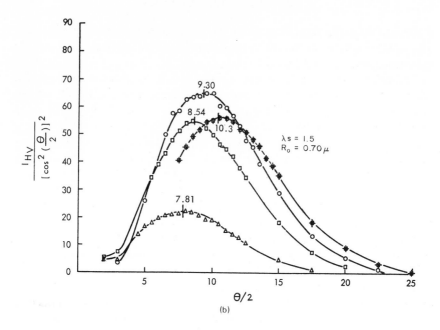

(b)

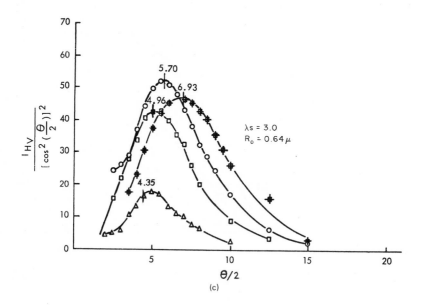

(c)

consistent with the known morphological characteristics of the spherulite. Therefore, the deformation of spherulites relative to the imposed uniaxial deformation of the film under different conditions of temperature and stress can be studied quantitatively. Similarly, deviations from the predicted form of the SALS pattern could yield valuable information about complex changes occurring in the spherulites. Such changes would be expected at large uniaxial deformations where disruptive processes are known to occur within the spherulites. Certainly, small-angle light scattering is an important new technique for nondestructively evaluating spherulite deformation processes in polymer films (and fibers) under controlled conditions of temperature and stress.

D. SUMMARY

In Chapter 2 an attempt has been made to demonstrate a unified, comprehensive approach to the characterization of the deformation of a polycrystalline polymer. To do so, the different morphological levels of the polymer were defined as gross sample deformation (characterized by the sample extension ratio), spherulite deformation, and molecular orientation. The objective was to be able to relate morphological information obtained from numerous independent experimental techniques at different morphological levels in a simple fashion.

To simplify the problem of attaining this objective, a single set of isotactic polypropylene samples that had all undergone a uniform, uniaxial deformation was used as the model system. The characteristics of these samples were then examined at the different morphological levels. The molecular level was considered first. A two-phase model of a polycrystalline polymer was proposed and the important morphological parameters were shown to be the fraction of crystals present in the sample, the intrinsic properties of the crystalline and noncrystalline regions of the polymer, and the orientation of each of these regions. This information alone acted as a unifying thread with which to relate morphological data obtained from x-ray diffraction, sonic modulus, birefringence, density, and infrared dichroism measurements, and led to new combinations of these techniques for the rapid acquisition of morphological data. It further developed insight into the significance of the observed information from these different methods so that fundamental molecular parameters such as the intrinsic lateral moduli and the intrinsic birefringences of the crystalline and noncrystalline regions, the principal polarizability difference per monomer unit, and the infrared transition moment angles could be obtained for the isotactic polypropylene samples.

Spherulite deformation and its relation to gross sample deformation was then considered. By relating the molecular orientation of the crystalline region (the crystal orientation function f_c) to the gross sample deformation

(λ), the spherulites were shown to be deformed affinely. In addition, the observed small-angle light scattering (SALS) from the deformed spherulites could be used to give quantitative information about the size of the original, undeformed spherulites and the deformation the spherulite had undergone. Obviously, for a nonaffine deformation, the SALS technique will similarly indicate how much deformation the spherulite has undergone in comparison with the different deformation of the gross sample.

Thus, by using a unified, comprehensive approach to the characterization of the deformation of polycrystalline polymer films, molecular, spherulitic, and gross sample deformation characteristics have been related. With the insights and information that have been made available, this approach can now be extended, in Chapter 3, to more complicated deformation processes.

References

1. R. Bonart, R. Hosemann, F. Motzkus, and H. Ruck, *Norelco Reptr.*, **7**, 81 (1960).
2. W. Ruland, *Polymer*, **5**, 89 (1964).
3. W. Ruland, *Acta Cryst.*, **14**, 1180 (1961).
4. G. M. Brauer and E. Horowitz, in *Analytical Chemistry of Polymers III*, G. M. Kline, Ed., Interscience, New York, 1962, Chap. 1.
5. W. O. Statton, in *Handbook of X-rays*, E. F. Kaelble, Ed., McGraw-Hill, New York, 1967, Chap. 21.
6. V. D. Gupta and R. B. Beevers, in *The Encyclopedia of X-rays and Gamma Rays*, G. L. Clark, Ed., Reinhold, New York, 1963, pp. 783–789.
7. R. Zbinden, *Infrared Spectroscopy of High Polymers*, Academic, New York, 1964, Chap. 1.
8. I. Y. Slonim, *Russ. Chem. Rev. English Transl.*, **31**, 308 (1962).
9. P. H. Hermans, *Contribution to the Physics of Cellulose Fibres*, Elsevier, New York, 1946.
10. C. W. Bunn, *Trans. Faraday Soc.*, **35**, 482 (1939).
11. H. P. Klug and L. E. Alexander, *X-Ray Diffraction Procedures*, Wiley, New York, 1954.
12. J. J. Hermans, P. H. Hermans, D. Vermaas, and A. Weidinger, *Rec. Trav. Chim.*, **65**, 427 (1946).
13. R. S. Stein, *J. Polymer Sci.*, **31**, 327 (1958).
14. Z. W. Wilchinsky, *J. Appl. Phys.*, **30**, 792 (1959).
15. R. A. Sack, *J. Polymer Sci.*, **54**, 543 (1961).
16. G. Natta, *Nuovo Cimento*, **15**, Series 10, Suppl., 40 (1960).
17. Z. W. Wilchinsky, *J. Appl. Phys.*, **31**, 1969 (1960).
18. R. J. Samuels, *Norelco Reptr.*, **10**, 101 (1963).
19. N. S. Gingrich, *Rev. Mod. Phys.*, **15**, 90 (1943).
20. R. S. Stein, J. Powers, and S. Hoshino, *Office Naval Res. Tech. Rept.*, No. 33, July 7, 1961.

21. The automatic system was obtained from General Electric Company for control of their XRD-5 diffractometer.

22. M. Polanyi, *Z. Physik*, **7**, 149 (1921).

23. S. Hoshino, J. Powers, D. G. Legrand, H. Kawai, and R. S. Stein, *J. Polymer Sci.*, **58**, 185 (1962).

24. R. S. Schotland, Ph.D. thesis, Polytechnic Institute of Brooklyn, Brooklyn, N.Y., 1962.

25. R. J. Samuels, *J. Polymer Sci.*, **A3**, 1741 (1965).

26. R. J. Samuels, *J. Polymer Sci.*, **C 20**, 253 (1967).

27. R. J. Samuels, *J. Polymer Sci.*, **A-2, 6**, 1101 (1968).

28. R. J. Samuels, *J. Polymer Sci.*, **A-2, 6**, 2021 (1968).

29. G. Farrow and J. Bagley, *Textile Res. J.*, **32**, 587 (1962).

30. J. H. Dumbleton, *J. Polymer Sci.*, **A-2, 6**, 795 (1968).

31. J. H. Dumbleton, *J. Polymer Sci.*, **A-2, 7**, 667 (1969).

32. J. H. Dumbleton, *Polymer*, **10**, 539 (1969).

33. R. J. Samuels, *J. Polymer Sci.*, **A-2, 7**, 1197 (1969).

34. M. Compostella, A. Coen, and F. Bertinotti, *Angew. Chem.*, **74**, 618 (1962).

35. M. E. Millberg, *J. Appl. Phys.*, **33**, 1766 (1962).

36. I. M. Ward, *Proc. Phys. Soc. London*, **80**, 1176 (1962).

37. I. M. Ward, *Textile Res. J.*, **34**, 806 (1964).

38. W. W. Moseley, Jr., *J. Appl. Polymer Sci.*, **3**, 266 (1960).

39. W. H. Church and W. W. Moseley, Jr., *Textile Res. J.*, **29**, 525 (1959).

40. S. E. Ross, *Textile Res. J.*, **34**, 565 (1964).

41. R. J. Urick, *J. Appl. Phys.*, **18**, 983 (1947).

42. H. A. Waterman, *Kolloid Z. Z. Polymere*, **192**, 9 (1963).

43. *Encyclopaedic Dictionary of Physics*, Vol. 2, Pergamon Press, New York, 1961, p. 21.

44. H. A. Waterman, *Kolloid Z. Z. Polymere*, **192**, 1 (1963).

45. J. W. Ballou and S. Silverman, *Textile Res. J.*, **14**, 282 (1944).

46. G. Oster and M. Yamamoto, *Chem. Rev.*, **63**, 257(1963).

47. I. Sakurada, T. Ito, and K. Nakamae, *Kobunshi Kagaku*, **21**, 197 (1964).

48. I. Sakurada, T. Ito, and K. Nakamae, *J. Japan Soc. Testing Mat.*, **11**, 683 (1962).

49. M. Horio, *Symposium at Polytechnic Institute of Brooklyn*, Sept. 7, 1963.

50. O. Wiener, *Abhandl. Saechs. Akad. Wiss. Leipzig, Math. Physik. Kl.*, **32**, 507 (1912).

51. F. A. Bettelheim and R. S. Stein, *J. Polymer Sci.*, **27**, 567 (1958).

52. R. S. Stein and F. H. Norris, *J. Polymer Sci.*, **21**, 381 (1956).

53. C. W. Bunn and R. de P. Daubeny, *Trans. Faraday Soc.*, **50**, 1173 (1954).

54. E. M. Chamot and C. W. Mason, *Handbook of Chemical Microscopy*, Vol. 1, Wiley, New York, 1951.

55. R. S. Stein, S. Onogi, and D. A. Keedy, *J. Polymer Sci.*, **57**, 801 (1962).

56. H. De Vries, *On the Elastic and Optical Properties of Cellulose Fibers*, Sehotanus and Jeus, Utrecht, 1953.

57. H. M. Morgan, *Textile Res. J.*, **32**, 866 (1962).

58. G. Farrow and J. Bagley, *Textile Res. J.*, **32**, 587 (1962).

59. R. S. Stein, *J. Polymer Sci.*, *A-2*, **7**, 1021 (1969).
60. M. F. Vuks, *Opt. i Spektroskopiya*, **2**, 494 (1957).
61. M. Peraldo, *Gazz. Chim. Ital.*, **89**, 798 (1959).
62. H. Tadokoro, M. Kobayashi, M. Ukita, K. Yasufuku, S. Murahashi, and T. Torii, *J. Chem. Phys.*, **42**, 1432 (1965).
63. D. A. Keedy, J. Powers, and R. S. Stein, *J. Appl. Phys.*, **31**, 1911 (1960).
64. K. G. Denbigh, *Trans. Faraday Soc.*, **36**, 936 (1940).
65. M. V. Volkenstein, *Configurational Statistics of Polymeric Chains*, Interscience, New York, 1963, Chap. 7.
66. V. N. Tsvetkov, in *Newer Methods of Polymer Characterization*, B. Ke, Ed., Interscience, New York, 1964, Chap. 14.
67. H. A. Szymanski, *Infrared Theory and Practice of Infrared Spectroscopy*, Plenum Press, New York, 1964.
68. G. W. King, *Spectroscopy and Molecular Structure*, Holt, Rinehart and Winston, New York, 1964.
69. S. Krimm, *Fortschr. Hochpolymer.-Forsch.*, **2**, 51 (1960).
70. C. Y. Liang, in *Newer Methods of Polymer Characterization*, B. Ke, Ed., Interscience, New York, 1964, Chap. 2.
71. G. Herzberg, *Infrared and Raman Spectra of Polyatomic Molecules*, D. Van Nostrand, New York, 1945.
72. H. C. Urey and C. A. Bradley, *Phys. Rev.*, **38**, 1969 (1931).
73. R. D. B. Fraser, in *Analytical Methods of Protein Chemistry*, Vol. 2, P. Alexander and R. J. Block, Eds., Pergamon, New York, 1960, Chap. 9.
74. R. D. B. Fraser, *J. Chem. Phys.*, **21**, 1511 (1953).
75. E. M. Bradbury, A. Elliot, and R. D. B. Fraser, *Trans. Faraday Soc.*, **56**, 1117 (1960).
76. R. D. B. Fraser, *J. Chem. Phys.*, **29**, 1428 (1958).
77. M. Tsuboi, *J. Polymer Sci.*, **59**, 139 (1962).
78. M. P. McDonald and I. M. Ward, *Polymer*, **2**, 341 (1961).
79. T. Miyazawa and Y. Ideguchi, *Bull. Chem. Soc. Japan*, **36**, 1125 (1963).
80. T. Miyazawa, *J. Polymer Sci.*, *C*, **7**, 59 (1964).
81. R. G. Snyder and J. H. Schachtschneider, *Spectrochim. Acta*, **20**, 853 (1964); *J. Polymer Sci.*, *C* **7**, 85 (1964).
82. S. Krimm, *J. Polymer Sci.*, *C* **7**, 59 (1964).
83. A. Keller, *J. Polymer Sci.*, **17**, 351 (1955).
84. H. D. Keith, F. J. Padden, Jr., N. M. Walter, and H. W. Wyckoff, *J. Appl. Phys.*, **30**, 1485 (1959).
85. R. J. Samuels and R. Y. Yee, *J. Polymer Sci.*, *A-2*, **10**, 385 (1972).
86. P. Ingram and A. Peterlin, *Polymer Letters*, **2**, 739 (1964).
87. V. I. Selikhova, G. S. Markova, and V. A. Kargin, *Vysokomolekul. Soedin.*, **6**, 1136 (1964).
88. S. Ishikawa, T. Kurita, and E. Suzuki, *J. Polymer Sci.*, *A* **2**, 2349 (1964).
89. F. J. Padden, Jr. and H. D. Keith, *J. Appl. Phys.*, **30**, 1479 (1959).
90. P. H. Geil, *Polymer Single Crystals*, Interscience, New York, 1963, Chap. 4.
91. M. Takayanagi, S. Minami, and H. Nagatoshi, *Asahi Garasii Kogyo Gijutsu Shoreikai*, **7**, 127 (1961).

92. Z. W. Wilchinsky, *Polymer*, **5**, 271 (1964).

93. R. S. Stein, M. B. Rhodes, P. R. Wilson, and S. N. Stidham, *Pure Appl. Chem.*, **4**, 219 (1962).

94. V. I. Selikhova, G. S. Markova, and V. A. Kargin, *Vysokomolekul. Soedin.*, **6**, 1132 (1964).

95. V. A. Kargin, T. D. Sogolova, and L. A. Nadareishvili, *Vysokomolekul. Soedin.*, **6**, 1407 (1964).

96. C. J. Speerschneider and C. H. Li, *J. Appl. Phys.*, **33**, 1871 (1962).

97. D. C. Bassett, *Phil. Mag.*, **6**, 1053 (1961).

98. J. A. Rusnock and D. Hansen, *J. Polymer Sci.*, *A* **3**, 647 (1965).

99. G. A. Bassett and A. Keller, *Phil. Mag.*, **9**, 817 (1964).

100. A. Guinier, G. Fournet, C. B. Walker, and K. L. Yudowitch, *Small Angle Scattering of X-Rays*, Wiley, New York, 1955.

101. R. J. Samuels, *J. Polymer Sci.*, *C* **13**, 37 (1966).

102. R. S. Stein and M. B. Rhodes, *J. Appl. Phys.*, **31**, 1873 (1960).

103. R. J. Samuels, *J. Polymer Sci.*, *A-2*, **9**, 2165 (1971).

104. L. C. Roess and C. G. Shull, *J. Appl. Phys.*, **18**, 308 (1947).

105. A. L. Patterson, *Phys. Rev.*, **56**, 972 (1939).

106. A. Guinier, *Ann. Phys. Paris*, **12**, 161 (1939).

107. R. S. Stein and J. J. van Aartsen, private communication.

3

Interpretation: Structural Interpretation of Fabrication Processes

A. INTRODUCTION

The purpose of Chapter 3 is to show how the structural parameters developed in Chapter 2 are applied to fabrication-process interpretation. The Series C film data used throughout Chapter 2 is used in this Chapter in conjunction with new data from other film processes (Series A and B films). Two fiber processes are also analyzed for their structural differences (Series D and E fibers). In this way the data from Chapter 2 are integrated into a much larger body of structural information from fibers and films, and the effect of process path variables on structural-state modification is illustrated.

It should be remembered that the fabrication process is simply a path to a given final structural state of the polymer. This section describes the effect of such path variables as temperature, draw ratio, process equipment, and sample form, on the final structure produced. In Chapter 4 the concept of equivalent structural states is developed, which shows that there are many different fabrication paths to a given structural state.

B. ISOTACTIC POLYPROPYLENE FILM FABRICATION PROCESSES

1. Introduction

Severe experimental and theoretical problems are encountered during any attempt to elucidate the deformation behavior of a polycrystalline polymer film. The film is a complex morphological structure composed of supramolecular spherulites and their connecting interspherulitic boundaries. The spherulites are composed of an organized substructure of crystalline lamellae and noncrystalline polymer. The crystalline lamellae have a preferred orientation with respect to the spherulite radii and are connected through tie molecules.

When such a complex structure is deformed the different structural entities can relieve the imposed stress by several deformation processes. Initially, the spherulites deform and, depending on the strength of the spherulite substructure, the interspherulitic boundaries may yield. Once the spherulite has deformed, the organized substructure will rearrange to relieve the imposed stress. The manner in which the rearrangement occurs will depend on the mechanical strength of the crystalline and noncrystalline components, the organizational character of the substructure, which dictates the manner in which the stress propagates, and the temperature and time of the deformation. Quantitative elucidation of such complex deformation processes must involve examination of the material at all of the morphological levels of behavior: the molecular, the interlamellar, and the spherulitic. Such a comprehensive characterization of the deformation behavior of uniaxially drawn isotactic polypropylene films necessitates the use of all the experimental techniques including sonic modulus, small- and wide-angle x-ray diffraction, birefringence, density, small-angle light scattering, electron microscopy, and optical microscopy discussed in Chapter 2.

The particular film processes examined here are represented by three series of drawn films. Films of different thicknesses were first cast on a roll at 30°C. The cast film thicknesses were varied so that all films would have the same final thickness (approximately 1.5 mils) after being drawn to different extensions. The films designated throughout this Chapter as Series A and as Series B were drawn at 110 and 135°C, respectively (1). The polymer used in the Series A and Series B films was Pro-fax 651A, a film-grade resin (Hercules) with a viscosity-average molecular weight of approximately 280,000 containing 3–4% Decalin-soluble material. The Series C films were also drawn at 110°C, but were made from a different polymer lot, a Pro-fax 6320 of $\overline{M}_v \simeq 190,000$ and containing 4–5% Decalin-soluble material (2). The Series C film was cast from the melt (ca. 280°C) on a 20°C roll at approximately 5 ft/min. The drawn films were produced by transferring cast film to a 110°C roll and hot drawing over a 5-ft gap to a room temperature roll. Sections of each of the films about 12 × 8 in. were annealed at fixed length for 15 min at 110°C and allowed to age for at least two months before being examined. It is the Series C films that were used in Chapter 2 to illustrate the different morphological techniques.

2. Molecular Anisotropy

When a polymer film is deformed, the crystalline and noncrystalline components orient to relieve the stress. As both the isotactic polypropylene crystal and the helical noncrystalline molecule are anisotropic, their orientation leads to anisotropic properties in the film. The important properties required to characterize the molecular anisotropy are the fraction of crystals

β, the average orientations of the crystal axes relative to the deformation direction, f_c and f_b, and the average orientation of the noncrystalline helical chain axis relative to the deformation direction f_{am}.

The fractions of crystals for the Series A and B films were obtained from density measurements, and the orientations of the c axis (helical chain axis) and b axis of the crystals were obtained from wide-angle x-ray diffraction measurements (Table 3-1). The values for the Series C films are the same as those reported earlier in Chapter 2, Tables 2-2 and 2-3.

The orientation of a crystal axis is represented numerically in terms of the Hermans orientation function:

$$f_x = \frac{3 \overline{\cos^2 \theta_x} - 1}{2} \tag{2-9}$$

When the x crystal axis is parallel to the deformation direction, f_x has a value of $+1.0$. When the x crystal axis is randomly oriented in the film, f_x is zero. When the x crystal axis is perpendicular to the deformation direction, f_x equals -0.5. The orientation functions for the three crystal axes are interrelated through the expression, $f_{a'} + f_b + f_c = 0$,* so that determination of two axes will completely characterize all three. The orientation of the crystals in the drawn films can, therefore, be represented by the orientation function triangle diagram illustrated in Fig. 3-1 (Chapter 2, Section B.2.d). A point at an apex of the triangle indicates the crystal axis is parallel to the deformation direction. A point at a side of the triangle indicates that the crystal axis is perpendicular to the deformation direction, and the midpoint 0, indicates that all the axes are randomly oriented. Since the original, undrawn films were unoriented, they are represented by the midpoint in Fig. 3-1. As the films are drawn, the c axis of the crystal orients toward the draw direction, and thus a movement from the midpoint toward the $c_{\parallel}$ apex of the triangle is a movement in the direction of increasing extensions of the film. The dashed straight line in the figure represents the condition in which the a and b axes of the crystal orient perpendicular to the deformation at the same rate, while the c axis orients parallel to the deformation direction, that is, the line is equidistant from both sides of the triangle. This is the case for cold-drawn polypropylene (3).

The points in Fig. 3-1 for the Series A, B, and C films do not follow the cold-draw line but bow toward the $b_{\perp}$ side of the triangle. This indicates that

---

* Owing to the monoclinic unit cell structure of isotactic polypropylene, the orientation function $f_{a'}$ is a measure of the orientation of a crystallographic direction mutually perpendicular to the b and c axes in the crystal, and not the true a-axis direction. This axis is designated as the a' axis and it is related to the true a axis through the deviation of the β monoclinic angle from $90°$. The β angle equals $99°20'$ for isotactic polypropylene and thus the a' axis orientation is close to the true a-axis orientation. In the discussion of the orientation triangle (Fig. 3-1) $f_{a'}$ is assumed, for clarity as an approximation for f_a (the true a-axis orientation function).

Table 3-1. Morphological Data for Isotactic Polypropylene Films

Sample	Drawing Temp., °C	Elongation, %	Density, g/cm³	β	Birefringence, $\Delta_T \times 10^3$	X-ray f_c	X-ray $-f_b$	Sonic Modulus $E_s \times 10^{-10}$, dyne/cm²	Sonic Modulus f_{am}	SAXS Drawn film L, Å	SAXS Preheat film L_0, Å
A-1	110	12	0.9030	0.604	4.413	0.3806	0.2726	3.065	−0.0571	138.5	133.8
A-2	110	32	0.9036	0.611	6.447	0.3946	0.2858	3.227	−0.0011	138.5	131.3
A-3	110	73	0.9035	0.610	9.170	0.4118	0.2818	3.585	+0.1113	138.5	130.9
A-4	110	162	0.9035	0.610	15.37	0.6495	0.3908	4.547	+0.2511	144.1	134.3
A-5	110	246	0.9050	0.630	21.67	0.7669	0.4238	5.179	+0.2769	150.1	134.3
A-6	110	447	0.9030	0.604	29.21	0.8994	0.4649	7.608	+0.5135	170.1	134.3
A-7	110	635	0.9030	0.604	29.29	0.9168	0.4543	9.300	+0.6026	170.1	136.3
B-1	135	89	0.9109	0.703	3.841	0.2678	0.3041	3.386	−0.1169	169.5	158.8
B-2	135	188	0.9110	0.704	10.07	0.4488	0.3469	4.574	+0.1771	177.9	160.2
B-3	135	325	0.9088	0.677	18.28	0.6443	0.3914	5.615	+0.3231	180.4	160.2
B-4	135	477	0.9081	0.667	26.66	0.8796	0.4488	7.585	+0.4353	191.3	159.5
B-5	135	610	0.9087	0.675	29.04	0.8925	0.4497	9.275	+0.5326	196.3	158.1
B-6	135	815	0.9086	0.674	30.80	0.9549	0.4753	11.092	+0.5857	196.3	157.4

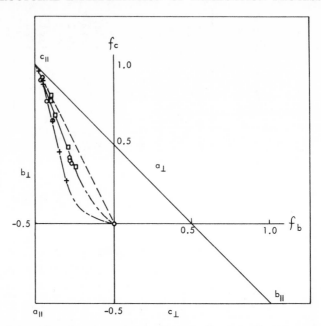

Fig. 3-1 Orientation function triangle diagram for isotactic polypropylene film: (O) Series A; (+) Series B; (□) Series C; (- - -) cold-drawn polypropylene (4); (-··-) extrapolation.

at low extensions the b axis of the crystal is orienting toward the perpendicular to the deformation direction faster than the a axis, and that the a axis does not catch up with the b axis until the chains are almost perfectly oriented in the deformation direction.

The Series A, B, and C films were drawn at elevated temperatures where crystal mobility is enhanced. Under these conditions homogeneous deformation of the spherulites occurs. In fact, the spherulites deformed affinely with the film (see Section B.4). The observed difference between the crystal orientation occurring at elevated temperatures and that occurring at room temperature is a direct consequence of the homogeneous character of the elevated temperature deformation. The crystallites in the polypropylene spherulite are oriented with the a axis of the crystal parallel to the axis of the radial fibrils. When the spherulite is deformed homogeneously some of the crystals rotate until their c axes are oriented parallel to the fibril-axis direction. This transformation occurs through a process of planar slip and rotation around the b axis of the crystal. The number of crystals undergoing this transformation is a function of the deformability of the spherulite (crystallinity effect), the perfection of the crystallites, and the temperature of the deformation.

The bow toward the $b_\perp$ side of the orientation function triangle diagram in Fig. 3-1 is simply an indication of the amount of a-axis orientation remaining in the sample. The greater the bow toward the $b_\perp$ side, the greater the effect of diffraction from a-axis-oriented crystallites on the c-axis-orientation function. The results in Fig. 3-1 indicate the higher the deformation temperature the smaller the number of a-axis-oriented crystals that convert to c-axis-oriented crystals. Thus, the Series B samples drawn at 135°C show more bow than the Series A and C samples drawn at 110°C. This is in agreement with earlier observations on Series C films and melt-spun fibers (Fig. 2-10). The results also indicate the crystallites in the 135°C samples are less deformable and more perfect and hence have greater difficulty in rotating about the b axis than those drawn at 110°C. These observations will be discussed in greater detail in the section on spherulite anisotropy (Section B.4).

The only information still required to describe the film anisotropy is the orientation of the molecules in the noncrystalline regions. This information was obtained for the Series A, B, and C films from both the sonic modulus (Chapter 2, Section B.3) and birefringence (Chapter 2, Section B.4) techniques. By applying the data in the tables to eqs. 2-35 and 2-36, f_{am} was calculated. The calculated f_{am} values are listed in Tables 2-3 and 3-1.

Molecular orientation in the Series A, B, and C films is illustrated in Fig. 3-2. Crystallite orientation in the Series A and C films (both deformed at the same temperature, 110°C) is the same. The solid lines designated f_λ and $f_{\lambda'}$ are defined as spherulite affine deformation limits (see Chapter 2, Section C.1) (4). The points from the Series A and C films fall within these limits, indicating that the spherulites in both these series deform affinely with the sample. The points from the Series A sample, which were corrected for shrinkage after machine drawing, fall directly on the $f_{\lambda'}$ line, indicating not only that the spherulites deformed affinely with the sample, but that the deformation mechanism is one that produces enhanced crystallite orientation in the inner regions of the spherulite. This type of deformation leads to radial curvature during extension (see the following section). Such a mechanistic interpretation is consistent with similar observations on low-density polyethylene (4). The Series B samples, drawn at 135°C, do not deform affinely, as the crystals orient more slowly with draw at the higher temperature. Nonaffine deformation in these films will be discussed below.

The orientation of the molecules in the noncrystalline region of the Series A and C samples is the same in the high-draw region but differs in the region below 100% extension. The lower molecular weight Series C samples show a greater tendency toward perpendicular orientation of the molecules than those of Series A (see Fig. 2-21). The Series B samples, drawn at higher temperature, show consistently lower f_{am} values over the whole deformation

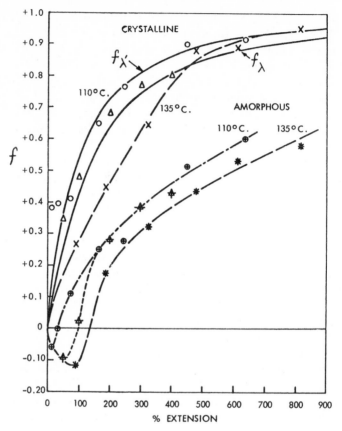

Fig. 3-2 Orientation functions for isotactic polypropylene films: (○) Series A; (×) Series B; (△) Series C; open symbols—Crystalline; crossed (+) symbols—noncrystalline.

range with a greater tendency toward perpendicular orientation in the low-deformation region than either of the lower-temperature drawn polymers. In fact, the overall orientation is consistently lower, for a given extension, for the samples drawn at 135°C than for those drawn at 110°C.

The observed average properties of the film are functions of the fraction of each phase present as well as of the orientation of the molecules in each phase. This is illustrated in the calculated contribution that the individual phases make to the total observed birefringence, shown in Fig. 3-3. Thus, because of the increased fraction of crystals in the samples drawn at 135°C, the crystals contribution in the high-draw region of these samples is greater than that of the samples drawn to equivalent extensions at 110°C. Similarly, the contribution the noncrystalline region makes to the total birefringence is affected by the fraction of noncrystalline polymer present. For example, not only is the orientation of the noncrystalline molecules in the films drawn at 135°C lower at all extensions than it is in the films drawn at 110°C, but the fraction of noncrystalline polymer is lower as well. This combination leads to a significantly smaller noncrystalline birefringence contribution from the films drawn at 135°C than from those drawn at 110°C. The observed lower total birefringence of the Series B films as compared with Series A and C films of similar extension is thus readily interpreted in terms of the combined molecular anisotropy contributions of the component regions of the film.

3. Interlamellar Anisotropy

The crystalline and noncrystalline regions develop into an ordered super-structure as a consequence of the spherulitic growth habit of isotactic poly-propylene films. The crystallites, in the form of folded chain lamellae, are

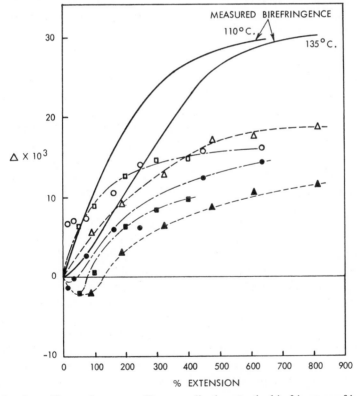

Fig. 3-3 Crystalline and noncrystalline contributions to the birefringence of isotactic polypropylene films: (○) Series A; (△) Series B; (□) Series C; open points—crystal contribution $[\beta\Delta_c^0 f_c]$; filled points—noncrystalline contribution $[(1 - \beta)\Delta_{am}^0 f_{am}]$.

arrayed along the radii of the spherulites. The average distance between their centers, L (which includes chain folds, tie molecules, noncrystallizable polymer, etc.) is on the order of 100–200 Å. This is just the size range that is most accessible to small-angle x-ray diffraction.

Just as wide-angle x-ray diffraction measures distances within the crystal (on the order of angstrom units), x-ray diffraction that occurs at very small angles, measures distances between the crystallites (on the order of a hundred angstroms). This is easily understood when the Bragg law is invoked (Chapter 2, eq. 2-6). Since $n\lambda = 2d \sin \theta$, $d = n\lambda/2 \sin \theta$. Here d is the crystal spacing (in small-angle x ray called the long spacing), λ is the wavelength of the x ray used, n is an integer, and θ is the Bragg reflection angle. According to the Bragg law, then, the smaller θ the larger will be the crystal spacing measured. Thus, at small angles, long spacings are measured.

Small-angle x-ray scattering (designated SAXS) is measured by both photographic and diffractometer techniques, just as with wide-angle x-ray diffraction (see Chapter 2, Section B.2.a). The small-angle x ray measures the distance between polymer crystal centers, which is in the range of hundreds of angstrom units. Most frequently the only cooperative reflections that usually appear in the SAXS patterns occur as a consequence of the periodic arrangement of crystal lamellae along the chain-axes (c-axis) direction. This is because there is considerable disorder in crystal matching in the other two directions from crystallite to crystallite. Since the long spacing measures the distance between crystal centers, it includes the noncrystalline polymer between the crystals in the c-axis direction as well. In fact, the intensity of the small-angle scattering is inversely proportional to the square of the electron-density difference between the crystalline and noncrystalline regions (e.g., $I_{SAXS} \approx (\rho_c - \rho_{am})^2$, where ρ_c is the electron density of the crystal and ρ_{am} is the electron density of the noncrystalline region).

Representative small-angle x-ray scattering (SAXS) patterns are shown in Fig. 3-4. The exposure time used to get these pictures varied since the scattering intensity decreased with increasing extension of the films. The spherulite deformation patterns are composed of a discrete reflection, which changes in both the distance of the reflection from the center of the pattern and the distribution of intensity around the center of the pattern as a function of the extension of the film. The distance of the discrete reflection from the center of the pattern is inversely proportional to the long spacing L, whereas the intensity distribution is a measure of lamellar orientation. The fiberlike deformation patterns are composed of a discrete meridian reflection that does not change in long spacing but increases in width along the layer line with extension.

The manner in which the long spacing L of the Series A, B, and C films is observed to change with extension is illustrated in Fig. 3-5 and listed in

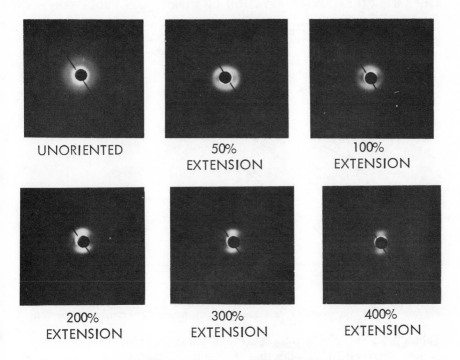

UNORIENTED   50% EXTENSION   100% EXTENSION

200% EXTENSION   300% EXTENSION   400% EXTENSION

SPHERULITE DEFORMATION
(LAMELLA ORIENTATION)

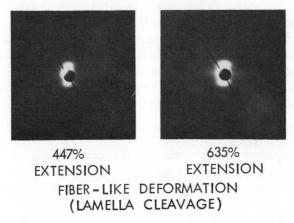

447% EXTENSION   635% EXTENSION

FIBER - LIKE DEFORMATION
(LAMELLA CLEAVAGE)

Fig. 3-4 SAXS patterns from isotactic polypropylene film.

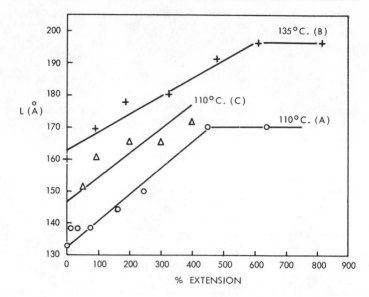

Fig. 3-5   SAXS long spacing L as a function of the percentage extension of the film.

Tables 3-1 and 3-2. The long spacing of the cast film used to prepare the Series C films was 128 Å, while that used for the Series A and B films averaged 108 Å. The difference in ordinate position of the curves for the Series A and C films (both of which were deformed at the same temperature) is primarily due to this crystallite size difference in the original cast films. The difference in the ordinate position of the curves for the Series A and B films is due to the difference in draw temperature; the higher temperature leads to greater annealing and thus to larger and more perfect crystallites. The temperature effect on crystallite size is further illustrated in Fig. 3-6 (Table 3-1). Here the long spacing L_0 of preheat samples of the cast films used to produce the Series A and B films is plotted against the cast-film thickness. The term preheat

Table 3-2. Long Spacings from Series C Isotactic Polypropylene Films

Sample	Elongation, %	Long Spacing, Å
C-2	50	151.4
C-3	100	160.6
C-4	200	165.4
C-5	300	165.4
C-6	400	171.8

signifies that the cast film was put through the same processing conditions as the Series A and B films except for the drawing operation. The long spacing L_0 for the undrawn preheat films is seen to be independent of the original cast film thickness and dependent only on the preheat temperature.

The most important features to notice in Fig. 3-5, however, are: (1) the long spacing increases linearly with extension at the two drawing temperatures up to some point, after which it stays constant with further extension and (2) the extension at which the long spacing becomes constant is different at the different draw temperatures ($\sim 450\%$ extension at $110°C$ and $\sim 600\%$ extension at $135°C$). Obviously, the mechanism of lamellar deformation in the region where the long spacing is changing with film extension must be different from that occurring at the high-extension region where the long spacing remains constant.

How can this complex SAXS behavior be explained? What type of deformation mechanisms can lead to these observations? The clues are present in the SAXS patterns, but it is necessary to consider the deformation morphology at all levels of structure, the molecular and spherulitic, as well as the interlamellar, in order to decipher them.

The undeformed films contain negatively birefringent spherulites (this was determined from small-angle light-scattering measurements on the films, see Section B.4) and, hence, the crystallites are initially oriented with their a axes parallel to the radial direction (5). This is shown schematically in Fig. 3-7. The radii in the undeformed spherulite have no preferred orientation,

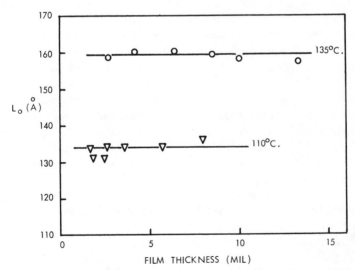

Fig. 3-6   Long spacing of preheat films as a function of film thickness: (∇) Series A; ($\bigcirc$) Series B.

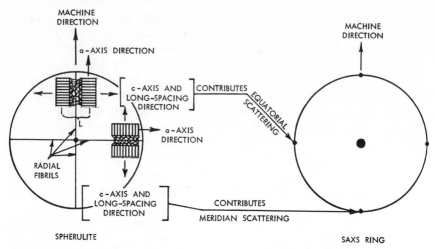

Fig. 3-7   Relation between intraspherulite lamellar placement and SAXS.

and thus the lamellae are spherically distributed symmetrically in space. The SAXS pattern from such an array will be a ring of constant azimuthal intensity with a 2θ radial position inversely proportional to the average interlamellar long spacing L of the sample (see Fig. 3-4, unoriented pattern). If the position of the radii within the spherulite are referred to a given direction in the film, that is, the machine or draw direction, then the position along the SAXS ring at which a specifically placed crystallite will contribute scattering can be identified. In the undeformed spherulite (see Fig. 3-7), lamellae aligned along radii parallel to the machine direction (in the polar region of the spherulite) will contribute SAXS scattering in the equatorial region of the SAXS pattern. This is a consequence of the a-axis orientation of the lamellae, which requires the long-spacing direction, that is, the same direction as the crystal c axis in the lamella, to be perpendicular to the machine direction. Similarly, lamellae aligned along radii that are perpendicular to the machine direction (in the equator of the spherulite) will contribute SAXS scattering in the meridian of the SAXS pattern. SAXS from lamellae aligned along intermediate radii will appear at intermediate positions between these extremes.

When the spherulite is deformed at a temperature high enough to allow crystal mobility, the c axis of the crystals, and hence the lamellar long-spacing direction, increasingly concentrates in the deformation direction. This is because the crystals are interconnected along their molecular axes (c axis) through tie molecules and any stress will tend to align this axis in the deformation direction. As the crystal lamellae orient preferentially in the deformation direction, a corresponding increase in the azimuthal intensity occurs in the meridian of the SAXS pattern (see Figs. 3-4 and 3-8). In fact, there is a continuous change from the ring of constant azimuthal intensity for the undeformed system to an arc for partial deformation and, finally, a spot on the meridian when fibrillation is complete and a line lattice of lamellae has been formed. Such a sequence of changes in the SAXS pattern is equivalent to similar changes in the wide-angle x-ray pattern, which suggests that lamellar orientation functions may be obtained from the SAXS patterns.

Figure 3-8 shows a model of the changes in lamellar placement within the spherulite that account for the observed SAXS changes during extension. Initial extension of the film causes a decrease in equatorial SAXS scattering, indicating that lamellar orientation along parallel radii has changed from a axis to tilted or c axis. Some a-axis orientation is observed in the wide-angle x-ray diffraction (see Fig. 3-1) throughout the spherulite deformation region of extension, however. Thus some of the a-axis-oriented lamellae along parallel radii must remain unchanged. As the stress would be expected to concentrate in the central region of the spherulite, initial extension must result in elongation of the near-central, parallel radii, with lamellae re-orientation and tilting, as the outer portion of the spherulite shows little change, besides a separation of lamellae along the equatorial fringe. The increasing intensity of meridional SAXS with a concurrent decrease in arc

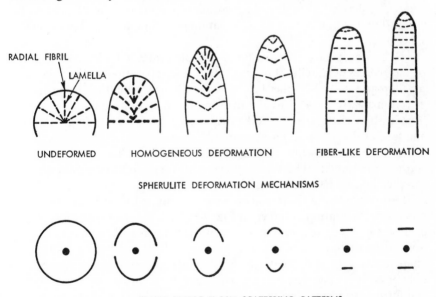

SPHERULITE DEFORMATION MECHANISMS

SMALL-ANGLE X-RAY SCATTERING PATTERNS

Fig. 3-8   Relation between intraspherulite deformation and the SAXS pattern.

length, and the diminishing amount of a-axis-oriented lamellae indicated by the wide-angle x-ray diffraction, show that as deformation continues, the size of the reoriented, central, parallel-radii portion increases at the expense of the a-axis-oriented, outer, parallel-radii region. This is required in order to preserve continuity within the spherulite as the outer portion of the spherulite in the region perpendicular to the deformation direction undergoes two types of deformation: (1) those off-axis radii subject to shear forces undergo lamellar slip and reorient with their long-spacing direction parallel to the deformation direction and (2) the distance between lamellae continues to increase with extension. Since lamellar separation is only one of the deformation processes occurring during extension, the change in long spacing is proportional, rather than equal, to the extension in Fig. 3-5.

A consequence of this complex structural reorientation is that a given radial line in the undeformed spherulite would become curved in the deformed spherulite. The observed changes in the crystal c-axis orientation function for the Series A and C films (Fig. 3-2) suggest that such radial curvature would be the case for these films, as the f_c values follow the theoretically derived $f_{\lambda'}$ curve. The assumptions for the derivation of the $f_{\lambda'}$ curve required curvature of a radial line with affine extension of the spherulite shape (4). The spherulite shape does change affinely with extension in these samples (see the following section). The proposed model is also in agreement with, and partly derived from, the observations of Hay and Keller (6) and Kobayashi (7) on the homogeneous deformation of two-dimensional polyethylene spherulites.

The homogeneous deformation of the spherulites will continue (Fig. 3-8) by the process of lamellar slip, orientation, and separation, until the lamellae in all regions of the spherulite become aligned with their long-spacing direction nearly parallel to the deformation direction. Once this stage of organization is reached, crystal orientation becomes exceedingly difficult, and further deformation must occur predominantly by crystal cleavage. At this point the transition from a spherulite to a fibrillar structure can be considered complete. The draw region in which crystal cleavage predominates can thus be characterized as the fiberlike deformation region (Fig. 3-8).

The dependence of the transition from spherulitic to fiberlike deformation on crystal orientation is shown in Fig. 3-9. Here the long spacing is plotted against the c-axis-orientation function. The long spacing increases with orientation up to the same limiting value $f_c = 0.90$, irrespective of the draw temperature. A crystal orientation function of 0.90 indicates crystals about as fully oriented as they can get. Beyond this value a line lattice is established, and further extension occurs by crystal cleavage, which leads to a constant long spacing with extension. The difference in the extension at which a constant long spacing occurs in the Series A and B films (Fig. 3-5) is simply a manifestation of the different rates of crystal orientation at the two draw temperatures, as shown in Fig. 3-2.

The noncrystalline region is also orienting during deformation (Fig. 3-2). This shows up in the intensity of the SAXS reflection, which decreases with increasing extension. As the SAXS intensity is proportional to the difference in electron density between the crystalline and noncrystalline regions, anything that decreases this difference will decrease the intensity. The greater the orientation of the noncrystalline region, the more crystallike it becomes and the smaller the electron density difference between the two regions. Hence, the decrease in SAXS intensity is a direct consequence of the increasing orientation of the noncrystalline region with extension. This again is in agreement with similar observations on polypropylene fibers (5).

Throughout the discussion the long-spacing direction of the lamellae has been considered to be the same as the c-axis direction in the crystallites, which, in turn, is the same as the molecular chain-axis direction in the crystal. As most of the crystals are in the form of lamellae aligned along radii within the spherulites, the orientation of the molecules within the crystallites should closely correspond to the orientation of the parent lamellae. Similarly, as the lamellae increasingly concentrate in the deformation direction with

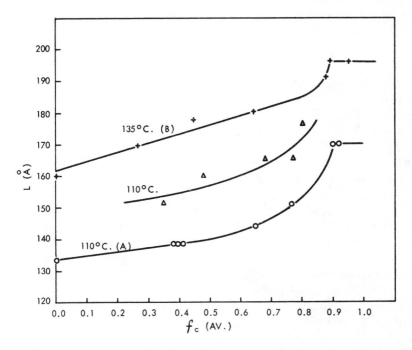

Fig. 3-9 Long spacing L as a function of the average c-axis orientation of the crystals.

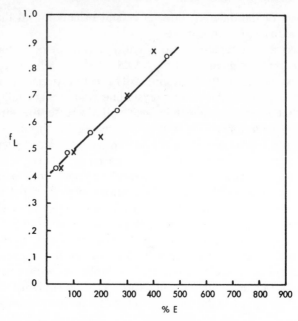

Fig. 3-10   SAXS orientation function from 110°C drawn isotactic polypropylene films: (O) Series A; (×) Series C.

increasing extension, their calculated orientation function should increase accordingly.

Lamellar orientation functions f_L were calculated from the angle of half-intensity of the SAXS diffraction arcs (8). Though this method is admittedly crude compared to the use of a diffractometer, it should serve as a good approximation to the true orientation behavior of the lamellae. Figure 3-10 is a plot of the lamellar orientation functions obtained from the Series A and C films as a function of the extension of the film. The lamellar orientation is found to increase continuously with increasing extension as is predicted by the model. The orientation functions for the Series B films could not be obtained because the discrete reflections were too close to the beam stop. Qualitative observation of the Series B SAXS patterns, however, indicated the orientation of the lamellae in samples B-1 to B-3 were less than would be expected from the extension ratio, in general agreement with the crystal orientation function data in Fig. 3-2.

The predicted correspondence between the orientation of the crystallites and the orientation of the lamellae is shown in Fig. 3-11. Here the lamellar orientation function for the Series A and C films is plotted against the orientation function for the molecular chain axis in the crystallites. The dashed line represents equal values for the two orientation functions. Considering the crudeness of the f_L measurement, an almost one-to-one correspondence is found between these two measurements. In light of recent suggestions (9) that the source of the SAXS reflection should be correlated with a morphological structure in the polymer only if there is independent support for the identification, this correspondence is gratifying.

The above SAXS results show that during high-temperature film extension there is a definite transition from homogeneous spherulite deformation, with lamellar orientation predominating, to fibrillar deformation where crystal cleavage predominates. The model in Fig. 3-8 suggests that in the fibrillar deformation region the lamellae are lined up along the fibrils like corn kernels along a cob with an interlamellar spacing in the fibril direction equal to the long spacing L. A SAXS measurement was taken of sample B-6 with the x-ray beam parallel to the machine direction in order to get an indication of the average lamellar dimension perpendicular to the fibril axis. Sample B-6 was chosen as it was the one with the highest crystallinity and the greatest crystallite orientation. The long spacing perpendicular to the fibril axis was found to be 167 Å. The long spacing in the fibril axis direction for this sample was 196 Å (Table 3-1). Thus, according to SAXS, the lamellae within the fibrils were rectangular blocks 196 Å high and 167 Å wide. As lamellae are often thought of as much wider than they are thick, a thorough electron-microscope examination of this film was made. Figures 3-12a and

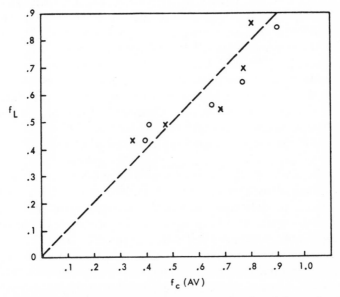

Fig. 3-11   Relation between the SAXS orientation function f_L and the c-axis orientation function f_c (av), from wide-angle x-ray diffraction: (○) Series A; (×) Series C.

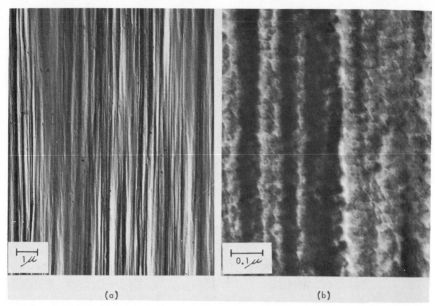

(a)   (b)

Fig. 3-12   Surface of highly drawn isotactic polypropylene films: (a) untreated (magnification 10,000 ×); (b) acid etched (magnification 150,000 ×).

3-12*b* are carbon surface replicas of sample 6-B. The film surface shown in Fig. 3-12*a* (magnified 10,000 times) is composed of fibrils 1000–2000 Å in diameter, which have a substructure of microfibrils. Figure 3-12*b* is the surface of the same film magnified 150,000 times after a 6-hr treatment with 70% nitric acid at 120°C (10). At this magnification the microfibrillar structure is clearly evident. The microfibrils are composed of essentially rectangular blocks aligned in an ordered array along the fibril. The blocks are 200–300 Å high and 200–500 Å wide. Considering the probable annealing effects of the severe acid treatment this is in good agreement with the SAXS results and further supports identification of the SAXS reflection with the lamellae.

The blocks in Fig. 3-12*b* are the crystalline lamellae, whereas the etched-out regions between them originally contained noncrystalline material. It is interesting to note that the weight loss from nitric acid treatment decreases with increasing film extension. This would be expected since increasing orientation of the noncrystalline region decreases the accessibility of the region to acid penetration. In agreement with the model proposed in Fig. 3-8, the fiberlike deformation region is thus characterized by highly oriented blocks of lamellae, aligned along the fibril axis in an ordered array, with oriented noncrystalline material between them.

The SAXS behavior developed for films in this section will be combined with the SAXS data obtained from fibers (discussed in Section C), to form general rules about crystallite deformation in Chapter 4, Section A.1.b.

4. Spherulite Anisotropy

A quantitative characterization of the spherulite type, size, and shape changes is needed to complete the morphological picture of film deformation processes. The small-angle light-scattering (SALS) technique was used almost exclusively to obtain this information, with the optical and electron microscopes acting as supplementary measurements.

The crystal form present in the Series A, B, and C films was monoclinic, as identified by x-ray diffraction, but the sign of the spherulite birefringence could not be determined with the optical microscope as the spherulites were too small. Consequently, an alternative route to the determination of the sign of the birefringence had to be explored. One possible route was through the small-angle light scattering (SALS) technique (see Chapter 2, Section C.2.a). Two types of SALS patterns can be obtained, an H_v SALS pattern and a V_v SALS pattern (11). A representative sample of these patterns, obtained from the cast and preheat films used to produce the Series A and B films, is shown in Fig. 3-13. Similar patterns were obtained from the unoriented Series C film (2). The H_v SALS pattern is obtained when the analyzer (which is placed between the sample and the photographic film) has its plane of polarization perpendicular to that of the polarized incident beam. This pattern is completely controlled by the shape factor in the H_v SALS equation and thus gives information about the size and shape of the spherulite (see discussion of shape anisotropy below). The V_v SALS pattern is obtained when the plane of polarization of the analyzer is parallel to that of the

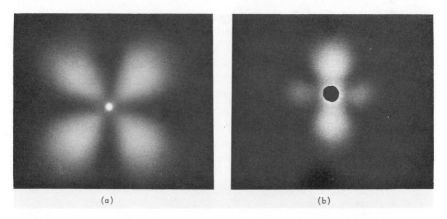

(a)   (b)

Fig. 3-13 Typical SALS patterns from isotactic polypropylene films (a) H_v SALS; (b) V_v SALS.

incident beam. The theoretical significance of this pattern is much more complex than that of the H_v SALS pattern, as it depends both on the spherulite anisotropy and on the difference between the spherulite polarizabilities and those of its surroundings. The spherulite polarizabilities are directly proportional to the refractive indices and it is the sensitivity of the polarizabilities to the spherulite surroundings in the V_v SALS pattern that opens a possible route to the determination of the sign of the birefringence of the spherulite. By examining the V_v SALS patterns from the films in a manner similar to that shown for isotactic polypropylene in Chapter 2, Section C.2.a and ref. 11 it was therefore possible to demonstrate that the spherulites were negatively birefringent. This information was important since it allowed a structural interpretation of the small-angle x-ray data to be made in Section B.3 from these same films.

The characteristic behavior of the H_v SALS pattern is determined by the form of the shape factor U. The theoretical equation describing the intensity of the H_v SALS for an anisotropic spheroid in an isotropic medium is (Chapter 2, Section C.2.a):

$$I_{H_v} = A\rho^2 V_0^2 \left(\frac{3}{U^2}\right)^2 [(\alpha_r - \alpha_t) \cos^2 \left(\frac{\theta}{2}\right) \sin \mu \cos \mu$$

$$\times (4 \sin U - U \cos U - 3 \operatorname{Si} U)]^2 \quad (2.68)$$

where the symbols are defined in Chapter 2, Section C.2.a. In an undeformed film the spherulite is spherical in shape and the shape factor takes the form:

$$U = \left(\frac{4\pi R_0}{\lambda'}\right) \sin \left(\frac{\theta}{2}\right) \quad (2.70)$$

Here, R_0 is the radius of the anisotropic sphere, λ' is the wavelength of light in the medium, and θ is the polar scattering angle. Owing to the dependence of the terms $(3/U^3)(4 \sin U - U \cos U - 3 \operatorname{Si} U)$ in the I_{H_v} equation on U, a maximum intensity will always be observed at a value of $U = 4.09$ provided $[\cos^2 (\theta/2)]^2$ is near unity. This means the average spherulite radius can be obtained from the H_v SALS pattern, as the distance from the center of the H_v SALS pattern to the intensity maxima of one of the lobes is, in conjunction with the known sample to film distance, a measure of the polar angle θ_{max}. Once the value of the polar angle has been obtained, the value of R_0 can be calculated from the equation:

$$R_0 = \frac{1.025\lambda'}{\pi \sin (\theta_{max}/2)}$$

The typical form of the SALS patterns obtained from the cast and preheat films used to prepare the Series A and B films is shown in Fig. 3-13.

Double SALS patterns were obtained from the cast and preheat films used to prepare samples B-4, B-5, and B-6 (Fig. 3-14), demonstrating that two size ranges of spherulites are present in these films. As the Series A and B films were designed to have approximately the same thickness after drawing, cast films of different thicknesses were required. The assumption was made that the spherulite size in the cast film depends only on the quench temperature and not on film thickness. Figure 3-15 is a plot of the average spherulite radius in the cast and preheat films, determined from the H_v SALS patterns, as a function of film thickness. These data show the following: (1) Spherulite size is independent of film thickness up to a thickness of 8 mils. This includes all of the cast and preheat films except those used to prepare samples B-4, B-5, and B-6. (2) Casting temperature, here 30°C, determines spherulite size, which remains constant during preheat. Preheat treatment leads to perfection of lamellae (Fig. 3-6). (3) A cast film of thickness greater than 8 mils leads to insufficient quenching with attendent internal heat retention. The thicker the film, the more inadequate the quenching and the greater the heat retention, yielding larger spherulite size. (4) As the films are quenched on a cold roll, inadequate quenching leads to two temperature zones in the film, one in contact with the roll, the other on the opposite side of the film; this results in two size ranges of spherulites in one film (Fig. 3-14).

The average radius R_0 of the spherulites in the films cast at 30°C was 0.59 μ. The value of R_0 for the spherulites in the Series C films cast at 20°C was 0.66 μ (2). The polymer used to make the Series C films had a lower molecular weight than that used to prepare the Series A and B films. The smaller spherulite size in the films cast at the higher temperature (30°C) is consistent with the smaller crystallite size found in the cast films, and the lower crystallinity of the Series A films (Table 3-1), and may be a result of the greater difficulty a higher molecular weight polymer has in crystallizing at a given temperature.

H_V V_V

Fig. 3-14   SALS double patterns from cast film.

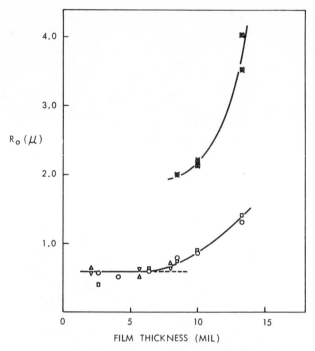

Fig. 3-15 The spherulite radius R_0 in different thickness cast and preheat films: (△) 110°C cast; (▽) 110°C preheat; (□, ⊠) 135°C cast; (O, ⊗) 135°C preheat.

The cast films were hot drawn at the prescribed temperature to form the Series A and B films. When a film is deformed the spherulites change their shape from a sphere to a prolate spheroid. The shape factor in the H_v SALS equation then has the form (12):

$$U = \left(\frac{4\pi R_0 \lambda_s^{-1/2}}{\lambda'}\right) \sin\left(\frac{\theta}{2}\right)\left[1 + (\lambda_s^3 - 1)\cos^2\left(\frac{\theta}{2}\right)\cos^2\mu\right]^{1/2} \quad (2\text{-}75)$$

where R_0 is the radius of the original, undeformed spherulite and λ_s is the extension ratio of the deformed spherulite. The factors that determine the form of the H_v SALS pattern during deformation are, thus, the size and extension of the spherulite.

A quantitative evaluation of the H_v SALS patterns from the Series C films showed that the spherulites were deforming affinely with the sample, that is, a given extension of the film led to an equivalent extension of the spherulites (see Chapter 2, Section C.2.b.1) (2,12). The H_v SALS patterns could not be obtained from the drawn Series A films as a result of the low crystallinity, small crystallites, and small spherulites, which together produced a deformed spherulite pattern too weak to extract from the diffuse background. The fact that the f_c, f_{am}, and SAXS behavior were the same for the Series A and C films, as well as the experimental verification of small spherulites in the cast and preheat Series A films (Figs. 3-13 and 3-15), all attest to the spherulitic nature and affine deformation behavior of the Series A films.

The deformation behavior of the spherulites in the Series B films is more complex than that of the other films. Any generalizations made about spherulite deformation at 135°C must depend on results obtained from samples B-1, B-2, and B-3, as samples B-4, B-5, and B-6 contain two size ranges of spherulites. The H_v SALS patterns obtained from the Series B films are shown in Fig. 3-16. The pattern obtained from film B-1 was too diffuse to reproduce clearly. The spherulite habits of the cast films is seen to be carried over into the deformed spherulite pattern. At low extensions the single H_v SALS pattern characteristic of a single size distribution of deformed spherulites is present, whereas at higher extensions the double deformed spherulite pattern predominates. At 815 % extension no spherulite pattern

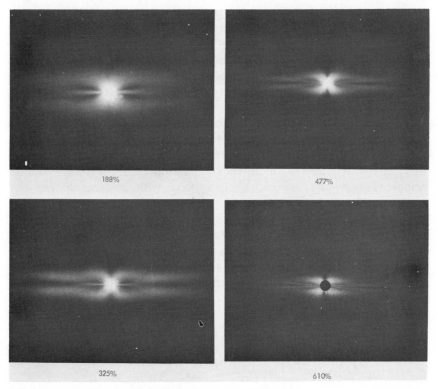

Fig. 3-16 H_v SALS patterns from films drawn at 135°C.

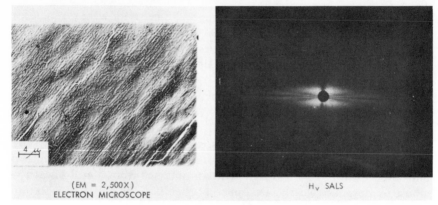

(EM = 2,500X)
ELECTRON MICROSCOPE

H$_v$ SALS

Fig. 3-17   Spherulites in deformed isotactic polypropylene film B5.

was observed because, as illustrated in Fig. 3-12, the film is highly fibrillar. In order to verify that the double H$_v$ SALS pattern is truly a result of the presence of two size ranges of spherulites in the film, a surface carbon replica was made of film B-5 and examined in the electron microscope. The replica of this film along with the corresponding H$_v$ SALS pattern is shown in Fig. 3-17. The information in the replica corresponds exactly to that in the H$_v$ SALS pattern, with a surface of highly oriented small spherulites over a subsurface of larger, less deformed spherulites clearly visible.

The extension ratio of the spherulites can be estimated from the H$_v$ SALS pattern by using the expression $\lambda_s = (h/d)^{2/3}$ where λ_s is the extension ratio of the spherulite, h is the measured distance along the length of a lobe, and d is the measured distance along the width of the same lobe (Chapter 2, Section C.2.b.1) (2). Calculation of the extension ratios of the spherulites in samples B-1, B-2, and B-3 indicates the spherulites deform affinely with the sample. This evidence is seemingly at variance with the observed deviation of the crystal orientation function f_c from the predicted affine deformation region in Wilchinsky's model (Fig. 3-2). The nonaffine character of the f_c results is obviously a consequence of the greater amount of a-axis orientation in these spherulites for a given extension than was found in the Series A and C samples (see Fig. 3-1). For example, at 200 % extension the Series A sample had 14 % a-axis-oriented material compared to 34 % for the Series B sample (5).

The greater amount of a-axis orientation during 135°C deformation indicates less disruption of the outer parallel radial region of the spherulite for a given extension at this temperature than at 110°C. This conclusion is further supported by the small change in lamellar orientation observed for samples B-1 through B-3 (see Chapter 3, Section B.3 on interlamellar anisotropy). Of course, the crystals are larger and more perfect at 135°C

(Fig. 3-6) and the amount of crystalline material is greater (Table 3-1); both factors tend to inhibit crystal deformation. The slower increase in long spacing with extension at 135°C (Fig. 3-5) also suggests less severe deformation of the system.

The lower f_{am} values of the Series B films (Fig. 3-2 and Table 3-1) reflect the greater mobility of the noncrystalline molecules at the higher temperature. This combination of less deformable crystals and greater mobility of the noncrystalline molecules results in an internal deformation mechanism which includes flow and relaxation of stress in the noncrystalline regions. Such a mechanism would lead to greater a-axis orientation of the crystals as the radii concentrate in the deformation direction with less reorientation of the crystallites and little disruption of the inner regions of the spherulite.

Wilchinsky's model (4) required the crystals to change their position and orientation in proportion to the deformation. Any situation in which some of the deformation is nonorientational in character must result in an observed nonaffine process. The nonaffine f_c behavior of the Series B films (Fig. 3-2) is thus a direct consequence of the nonorientational relaxation mechanisms in the noncrystalline region.

It is concluded from these observations that during very-high-temperature deformation the spherulite boundaries deform affinely with the sample, but the intraspherulite deformation mechanisms include nonorienting relaxation processes within the noncrystalline region.

5. Summary

The primary purpose of this examination was to show how the structural data are used to obtain a quantitative morphological description of the complex deformation processes occurring during high-temperature extension of spherulitic films. The use of a battery of independent experimental techniques has been successful in accomplishing this goal and has led to the following conclusions about the deformation mechanisms involved.

In all cases the films contained negative birefringent spherulites with the a axis of the lamellae within the spherulites aligned parallel to the radial direction. When such a film is deformed uniaxially at 110°C the spherulites deform affinely with the sample. Within a spherulite the deformation mechanism depends on the location of the radii with respect to the applied load. Along the radii aligned parallel to the extension direction, disruption and realignment of lamellae occur in the inner regions of the spherulite, whereas little or no change occurs in the outer spherulite regions. The lamellae aligned along radii in the region of the spherulite transverse to the deformation direction undergo two processes: (1) lamellar slip leading to c-axis orientation and (2) separation of lamellae. During these processes the helical chain axes of the noncrystalline molecules initially orient slightly more toward the perpendicular to the extension direction then random orientation, but with increasing extension they orient increasingly in the deformation direction.

This process of increasing c-axis orientation of the crystalline and non-crystalline regions continues until the lamellae become as fully aligned as they can without crystal cleavage processes predominating. Once this point is reached any further extension of the film results in a new deformation mechanism, crystal cleavage. In this highly extended state the substructure is no longer spherulitic but has evolved into a fibrillar structure. To accommodate further extension the lamellae cleave and break off in blocks and the helical axis of the noncrystalline molecules becomes more oriented in the deformation direction.

When films are drawn at 135°C, the deformation behavior is similar but not identical to that occurring during 110°C extension. The spherulites deform affinely with the film at this temperature and the transition from crystal orientation to crystal cleavage processes occurs when the chain-axis direction in the lamellae becomes fully oriented in the deformation direction, just as they did at 110°C. However, the lamellae are larger and more perfect at this high temperature and thus less deformable, and the noncrystalline molecules are more mobile. This leads to diminished internal disruption within the spherulite during extension, as relaxation processes within the noncrystalline region result in deformation processes that are nonorienting in character.

C. ISOTACTIC POLYPROPYLENE FIBER FABRICATION PROCESSES

1. Introduction

The preparation of isotactic polypropylene multifilament drawn yarn generally proceeds in three stages: (1) spinning of the fiber from a melt, (2) drawing the spun fiber, and (3) heat setting of the drawn fiber under tension. The organization of the substructure within the spun fiber at all morphological levels, the spherulitic, the interlamellar, and the molecular is complex. Its particular character will depend on the spinning conditions. Subsequent drawing of the spun fiber results in a reorganization of the substructure. The character of the reorganization is expected to depend not only on the environmental drawing conditions but on the original structure present in the spun fiber as well. A further rearrangement of the substructure of the drawn fiber occurs during the heat-setting stage of the process. Again, the nature of the rearrangement depends not only on the heat-setting environment but also on the substructure of the drawn fiber at the time of the setting operation. Each time a multifilament drawn yarn is produced, this type of complex structural reshuffling occurs. And each time the environmental conditions of spinning, drawing, and heat setting are changed, the details of the path and the structure of the final yarn are different.

A quantitative theory explaining the effect of rate, temperature, tension, and other environmental factors on the final structure of a drawn yarn cannot be derived until it is possible to describe in quantitative form the morphological changes that occur at different stages of the process. The same *modus operandi* used for characterizing film processes in the previous section is used in this section to characterize a fiber process in the desired quantitative morphological terms. The approach taken is analogous to stop-action photography. That is, a spun-fiber structure is quantitatively characterized at the molecular, interlamellar, and spherulitic levels (5). The spun fiber is then drawn (at different draw ratios), and each of the drawn fibers is quantitatively characterized in a similar manner. Finally, the process is completed as each of the drawn fibers is heat-set under tension and its structure characterized (5,13). In this way the fiber-forming process can be described, from spun fiber to final yarn, in quantitative morphological terms.

The different steps in the fiber process considered here are represented by two series of fibers. To prepare the Series D (drawn) fibers a normal spun polypropylene fiber (air quenched) was drawn to different draw ratios at 90°C and cooled under tension on a 42°C roll (this is the equilibrium temperature the unheated roll reached by taking heat from the yarn). All of the draw ratios reported for this series of fibers were calculated from denier reduction. The Series D fibers represent the drawing step in the fiber process.

To prepare the Series E (heat-set) fibers the same spun fiber was drawn to different draw ratios at 90°C and then was heat set under tension at 140°C. This high temperature was chosen to characterize effects resulting from extreme processing conditions. Again, all draw ratios reported were calculated from denier reduction (5). This series of fibers is the final product of the complete process.

2. Process Analysis

a. Spun Fiber. The spun fiber was produced by forcing a melt through a small die hole and keeping it under tension while it was air quenched. A combination of the normal forces in the swollen liquid at the die and the tension maintained during the air quench caused the resulting spun fiber to be oriented, as is attested by its high birefringence (Tables 3-3 and 3-4). This combination of forces also leads to the type of spherulite formation shown in Figs. 3-18 and 3-19. From the H_v SALS patterns (Figs. 3-29 and 3-30), the spun fiber is seen to be composed of 0.7-μ spherulites whose long axes are normal to the fiber-axis direction.

The internal structure of this spun fiber is quite complex (Fig. 3-20a). The

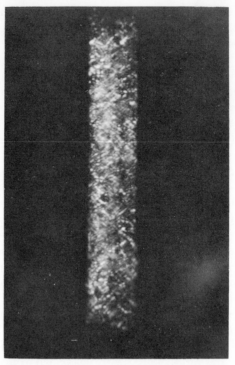

Fig. 3-18 Optical photomicrograph of spherulite structure within a spun fiber of isotactic polypropylene.

crystallites within the spherulites have a-axis orientation; those in the interspherulitic fibrillar links have c-axis orientation (14,15). If the fiber axis is chosen as the reference direction (Fig. 3-20b), then the crystallites along a spherulite radius parallel to the fiber axis will manifest a-axis orientation, while those along a radius perpendicular to the fiber axis will manifest c-axis orientation. If the spherulite had been homogeneously deformed, the crystallite orientation would be uniformly distributed between these extremes (16). However, this is not the case in the spun fiber. The distribution of crystal orientations, derived from the wide-angle x-ray scan (Figs. 3-21 and 3-25, see also Fig. 2-11), consists of two distinct regions, an a-axis-oriented region and a c-axis-oriented region, with a sharp boundary between. Such a distribution would result if yielding had occurred in the spherulites to produce an oriented boundary between the a-axis-oriented and c-axis-oriented regions. From the fraction of a-axis-oriented material this yield region would have occurred at 31.5° from the fiber-axis direction. The fraction of a-axis oriented crystals within the spun fiber was determined as follows. A representative bimodal azimuthal scan of a (110) plane is illustrated

MAGNIFICATION 2,500X   MAGNIFICATION 9,500X

Fig. 3-19 Surface replica of the spherulitic structure in isotactic polypropylene fiber as viewed in the electron microscope.

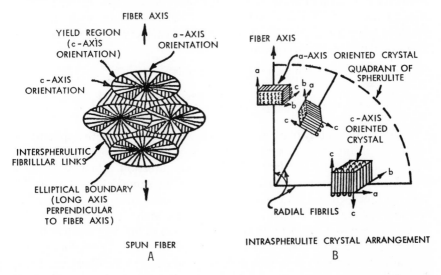

Fig. 3-20 Schematic representation of (A) spun fiber and (B) normal undeformed, spherulite morphologies.

143

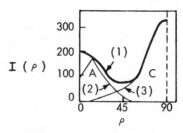

Fig. 3-21   Azimuthal scan of the 110 crystal reflection from spun fiber.

in Fig. 3-21, curve 1. To separate the bimodal curve into its two components the following procedure was used. It was assumed that the peak of the a-axis component was at $\rho = 14°$ and that one tail went to zero at $\rho = 55°$. Since the a-axis reflection was composed of two spots symmetrical about the meridian, half of the intensity at $\rho = 0°$ was assumed to be due to the a-axis spot in the working quadrant. Since this spot should have a symmetrical intensity distribution about its peak, this same half-intensity value was placed at $\rho = 28°$. The best curve was drawn between the points at $\rho = 14°$, 28°, and 55° and was made symmetrical through $\rho = 0°$ to give curve 2 in Fig. 3-2. The area under this a-axis-oriented component of the (110) reflection is designated A in the figure. Curve 2 was then subtracted from curve 1 between $\rho = 14°$ and $\rho = 90°$ to produce curve 3, the residual c-axis-oriented component of the scan. The area under curve 3 is designated C. The areas A and C were calculated by an SDS920 computer with $I(\rho)$ data obtained from curves 2 and 3. The fraction of each crystal component was then calculated from the equations:

$$\phi_A = \text{fraction of } a\text{-axis-oriented crystals} = \frac{A}{A + C}$$

$$\phi_c = \text{fraction of } c\text{-axis-oriented crystals} = (1 - \phi_A) = \frac{C}{A + C}$$

A deformation study of a microtomed section of bulk negative spherulites was made in order to substantiate the x-ray interpretation that the yield region occurred at 31.5° from the fiber-axis direction. These spherulites (Fig. 3-22) fortuitously underwent essentially the same deformation as did the spun-fiber spherulites ($\approx 20\%$). Yielding occurred in these spherulites between 40° and 80° from their long axis, the major yield being at 60°. This corresponds to a yield angle of 30° (range of 10–50°) relative to the fiber axis in the spun fiber, because the small, deformed spherulites in this fiber are oriented with their long axes perpendicular to the fiber axis. Thus, the yielding region is the same for the small spherulites in the spun fiber and the large, negative spherulites grown in bulk.

Thus the spun fiber is composed of inhomogeneously deformed spherulites and interspherulitic fibrillar links (Fig. 3-24). The deformed spherulite is composed of a yield region, c-axis-oriented, at an angle of 30° to the fiber axis, a region above the yield line, which has a-axis orientation with respect to the fiber axis, and a region below the yield line that is c-axis oriented. The interspherulitic fibrillar regions have c-axis orientation, whereas the non-crystalline regions (both interspherulitic and intraspherulitic) are slightly oriented in the fiber-axis direction (Fig. 3-24a). The question at issue is: how does such a complicated structure behave during the drawing and heat-setting stages of a fabrication process?

b. Deformation Processes. The specific fabrication process to be described is the drawing of the spun fiber at 90°C with subsequent heat setting under tension at 140°C. Crystals of isotactic polypropylene develop enhanced mobility at 110°C (the crystal absorption temperature of the dynamic

Fig. 3-22 Deformed bulk isotactic polypropylene spherulites: (a) microtomed section in polarizing microscope (20 × magnification).

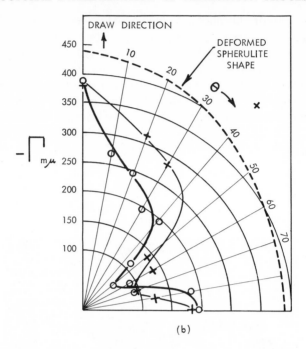

(b)

Fig. 3-22 (b) Retardation of deformed bulk spherulites as a function of azimuthal angle θ. Deformation direction vertical.

mechanical spectra) (17). Since the draw temperature (90°C) is below the crystal-mobility temperature (110°C), and the heat-setting temperature (140°C) is above the crystal-mobility temperature, distinct structural differences are expected at the two stages of the process. A characterization of the deformation the crystalline and noncrystalline regions underwent during the two stages of yarn fabrication was obtained from a combination of wide-angle x-ray diffraction, birefringence, and density measurements. The orientation of a crystal axis is represented numerically in terms of the Hermans' orientation functions, $f_x = (3 \overline{\cos^2 \theta_x} - 1)/2$ (see eq. 2-9). When the x axis is parallel to the deformation direction, f_x has a value of $+1.0$. When the x crystal axis is randomly oriented, $f_x = 0$. When the x crystal axis is perpendicular to the deformation direction, $f_x = -0.5$. The orientation functions for the three crystal axes are interrelated through the expression, $f_c + f_b + f_{a'} = 0$, so that determination of two axes will completely characterize all three. The orientation of the crystals in the fibers can, therefore, be represented by the orientation function triangle diagram illustrated in Figs. 2-10 and 3-23. The orientation function for the non-crystalline molecules f_{am} is calculated from a combination of the birefringence

Δ_T, the x-ray orientation function f_c, and the fraction β of crystals, determined from density measurements, and substitution of the value of these parameters into the equation $\Delta_T = \beta\Delta_c^0 f_c + (1 - \beta)\Delta_{am}^0 f_{am}$. Here Δ_c^0 and Δ_{am}^0 are the intrinsic birefringence of the crystal and noncrystalline molecule, respectively. The results of these measurements are illustrated in Figs. 3-23 to 3-26. Interlamellar behavior is characterized by small-angle x-ray measurements of the long spacing L between crystals in the chain-axis direction (Fig. 3-27) while spherulite deformation was followed with small-angle light scattering (Figs. 3-28 to 3-30). The data for the drawn series of fibers are given in Table 3-3, while those for the heat-set fibers are reported in Table 3-4. These results, as expected, show striking differences between the structural state of the spun fiber, the structure of the fiber after drawing to different draw ratios at 90°C (drawn series), and the structure of the final fabricated fibers after heat setting under tension at 140°C (heat-set series).

Changing stages in the structure of the fiber during fabrication, as described by the morphological data in Figs. 3-23 to 3-30 and Tables 3-3 and 3-4, are schematically represented in Fig. 3-31. In a negative polypropylene spherulite, the a-axis of the lamellae is oriented parallel to the radii of the spherulite (13). In the spun fiber, the spherulite is deformed with its long axis perpendicular to the fiber axis (Figs. 3-28, 3-29, spun fiber). There is an abnormal

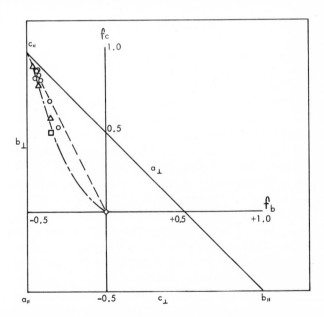

Fig. 3-23 Orientation function triangle diagram for the deformed fibers: (○) drawn; (△) heat set; (□) spun; (---) cold-drawn films (7).

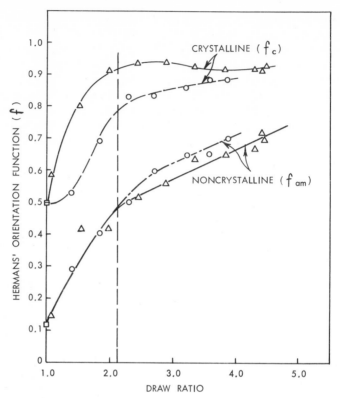

Fig. 3-24 Crystalline and noncrystalline orientation after deformation: (□) spun; (○) drawn; (△) heat set.

amount of a-axis- and c-axis-oriented lamellae with a deficiency of intermediate orientations within the spherulite (Fig. 3-25, spun fiber). This suggests a structure where the radii in the polar region of the spherulite (a region between the fiber axis direction and 30° from the fiber axis; see spun fiber, Fig. 3-31) are all curved toward the fiber axis, and the radii in the equatorial region of the spherulite (from 30° to 90° from the fiber-axis direction) have the normal radial character of an undeformed spherulite. This type of curved radii structure is similar, but not identical, to that obtained during high-temperature deformation of isotactic polypropylene spherulites on the assumption that the draw direction of the spherulites in the spun fiber is perpendicular to the fiber axis (Section C.2.a).

Two draw regions characterized by different deformation behavior can be identified from the structural data: the low-draw region, which occurs below the draw ratio range of $2.0–2.25\times$, and the high-draw region, from $2.0–2.25\times$ to the highest draw ratio. The ultimate draw ratio is limited by the structure developed, which in turn is determined by the fabrication conditions.

In the low-draw region, structural reorganization occurs in two stages: first during drawing, and then during heat setting under tension. During the drawing stage the spun fiber is cold drawn. This is shown in Fig. 3-23, where the data for the drawn series follow the dashed line characteristic of crystal orientation in cold-drawn isotactic polypropylene film (3). Cold drawing would be expected, since deformation occurs below the crystal mobility temperature. The lamellae at this temperature resist orientation and the molecules in the noncrystalline region orient most rapidly (Fig. 3-24).

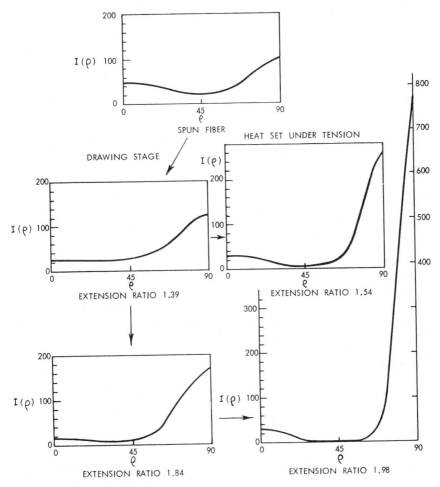

Fig. 3-25 Wide-angle x-ray diffraction (110) azimuthal intensity distributions.

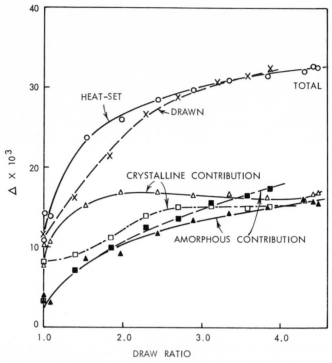

Fig. 3-26 Characterization of the birefringence contributions from the crystalline and noncrystalline regions during deformation: (□) drawn; (△) heat set; open points $[\beta \, \Delta_c^0 f_c]$; filled points $[(1 - \beta) \, \Delta_{am}^0 f_{am}]$.

A slight increase in crystallinity occurs due to annealing effects of the increased temperature, which also increases the long spacing (interlamellar spacing) to 134 Å. The long spacing increases with extension of the spun fiber, because of the controlling influence of the strained noncrystalline chains (Fig. 3-27). The spherulites are deformed with extension until their long axes are in the direction of the fiber axis (Fig. 3-28). Here λ_s is the extension ratio of the spherulite (2). A negative value for λ_s indicates the long axis of the spherulite is perpendicular to the fiber axis (the reference axis is perpendicular to the fiber axis when λ_s is defined as negative) and a positive value indicates the long axis of the spherulite is in the direction of the fiber axis (the reference axis is parallel to the fiber axis when λ_s is defined as positive). The spherulite extension is not affine so that some interspherulitic deformation occurs as well.

Spherulite deformation (Figs. 3-29 and 3-30) proceeds by disruption of the polar region. This occurs by a tilting of the a-axis-oriented lamellae (Fig. 3-25). These changes are illustrated in Fig. 3-31 (drawing stage). In the figure

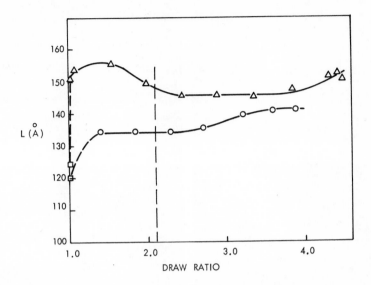

Fig. 3-27 Effect of draw ratio on the long spacing (L) of the fibers: ($\triangle$) heat set; ($\bigcirc$) drawn; ($\square$) spun fiber.

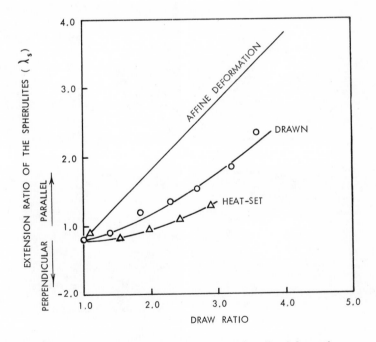

Fig. 3-28 Effect of temperature and draw ratio on spherulite deformation.

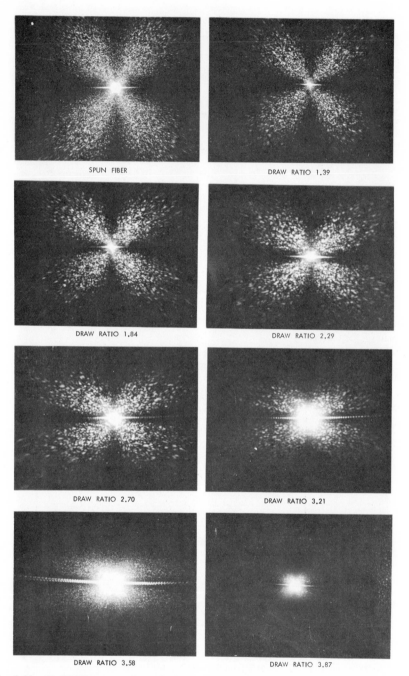

SPUN FIBER

DRAW RATIO 1.39

DRAW RATIO 1.84

DRAW RATIO 2.29

DRAW RATIO 2.70

DRAW RATIO 3.21

DRAW RATIO 3.58

DRAW RATIO 3.87

Fig. 3-29 H_v SALS patterns from the Series D (drawn) fibers. The polarization direction and the fiber-axis direction are vertical.

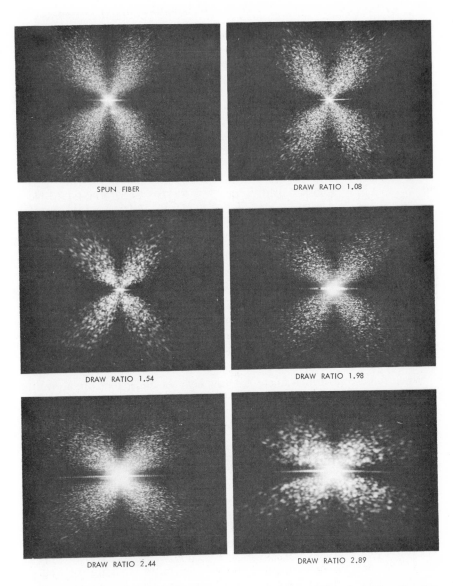

SPUN FIBER

DRAW RATIO 1.08

DRAW RATIO 1.54

DRAW RATIO 1.98

DRAW RATIO 2.44

DRAW RATIO 2.89

Fig. 3-30 H_v SALS patterns from the Series E (heat set) fibers. The polarization direction and the fiber-axis direction are vertical.

Table 3-3. Morphological Data for the Series D (Drawn) Isotactic Polypropylene Fibers

Sample	Draw Ratio	Density, g/cm^3	β	Bire-fringence, $\Delta_T \times 10^3$	Orientation Functions			SAXS, L, Å
					f_c	$-f_b$	$+f_{am}$	
Spun	1.00	0.9014	0.584	11.52	0.4996	0.3510	0.1247	120
1	1.39	0.9026	0.600	16.17	0.5289	0.3030	0.2883	134
2	1.84	0.9020	0.593	21.23	0.6933	0.3570	0.4000	134
3	2.29	0.9020	0.593	26.52	0.8285	0.4110	0.4976	134
4	2.70	0.9052	0.632	28.59	0.8324	0.4450	0.5978	135
5	3.21	0.9035	0.610	30.45	0.8573	0.4220	0.6451	140
6	3.58	0.9022	0.594	31.25	0.8810	0.4310	0.6510	141
7	3.87	0.9027	0.601	32.39	0.8829	0.4350	0.7002	141

the spherulite has been drawn out from an ellipse, with the long axis perpendicular to the fiber axis (spun fiber), into a sphere. This has been accomplished by a cold drawing of the central a-axis-oriented region of the spherulite, with the result that many of the lamellae in previously curved radii have tilted into a more normal orientation, with a consequent decrease in the amount of a-axis-oriented curved radii. The sharp boundary between the polar and equatorial regions of the spherulite has become less definite as a result.

When these cold-drawn fibers are now heat set under tension, above the crystal-mobility temperature, major changes occur in the equatorial region

Table 3-4. Morphological Data for the Series E (Heat-Set) Isotactic Polypropylene Fibers

Sample No.	Draw Ratio	Density, g/cm^3	β	Bire-fringence, $\Delta_T \times 10^3$	X-ray			SAXS, L, Å
					f_c (av)	$-f_\beta$	$+f_{am}$	
Spun	1.00	0.8994	0.558	10.87	0.4900	0.3549	0.1125	124.6
1	1.08	0.9064	0.650	13.93	0.5849	0.3477	0.1433	153.6
2	1.54	0.9083	0.668	23.74	0.7977	0.4265	0.4170	155.5
3	1.98	0.9067	0.650	25.95	0.9113	0.4638	0.4191	149.6
4	2.44	0.9052	0.632	28.46	0.9329	0.4684	0.5127	145.7
5	2.89	0.9039	0.615	29.70	0.9366	0.4690	0.5582	145.7
6	3.35	0.9055	0.635	30.96	0.9245	0.4572	0.6310	144.8
7	3.84	0.9045	0.624	31.44	0.9151	0.4460	0.6526	147.2
8	4.31	0.9033	0.610	31.99	0.9156	0.4477	0.6667	151.4
9	4.42	0.9058	0.640	32.60	0.9138	0.4527	0.7163	152.2
10	4.48	0.9061	0.642	32.52	0.9281	0.4563	0.7024	150.6

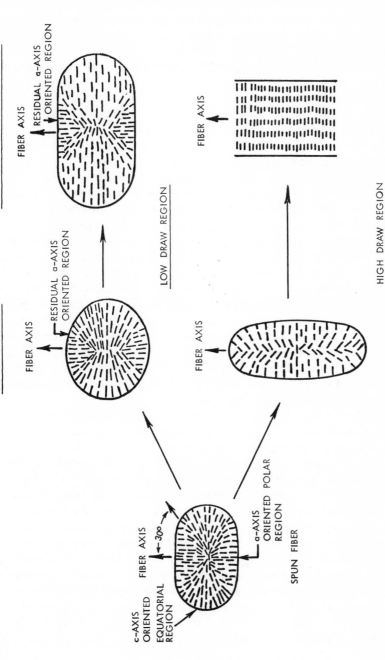

Fig. 3-31 Schematic representation of spherulite deformation mechanisms.

155

of the spherulites. The tilted lamellae, given thermal mobility, rotate so that their chain axes are oriented in the fiber-axis direction (c-axis orientation). Little or no change occurs in the remaining a-axis-oriented polar region, and a sharp boundary between the two regions is reformed. This is shown in Fig. 3-25, where the crystallite distribution for the heat-set fiber redevelops bimodal character with little change in the a-axis region compared to that of a fiber of a closely related draw ratio from the drawn-fiber series. The rotation of lamellae and an observed 5 % increase in crystallinity are the primary reasons for the observed increase in birefringence (Fig. 3-26) in the low-draw region, as the orientation of the molecules in the noncrystalline region does not change during annealing. The amount of noncrystalline polymer is decreased by the thermal treatment, however.

Annealing during heat setting raises the long spacing to 151 Å, and a further slight rise in long spacing is observed with increasing draw ratio. Very soon, however, the constraints of the strained noncrystalline region inhibit the rate of annealing and the long spacing begins to decrease.

Examination of the small-angle light-scattering patterns from the fibers heat set under tension (5) indicates that the extension ratio of the spherulite decreases slightly (Fig. 3-28). This most likely implies an extension of the spherulite in the direction perpendicular to the fiber axis, rather than a retraction of the polar region. The lamellae are lathlike with their greatest length in the a-axis direction and shortest length in the c-axis direction (18). Rotation of a large number of lamellae from tilted orientation to c-axis orientation would tend to increase the length of the spherulite in the direction perpendicular to the fiber axis, as observed. The transition from the drawn spherulite structure to that produced by heat setting under tension is schematically represented in Fig. 3-31.

Structural reorganization within the spherulite in the high-draw region, above a draw ratio of 2.0–2.25 ×, also occurs in two stages: (1) during drawing and (2) during heat setting under tension. As the spun fiber is cold drawn beyond a draw ratio of 2.0–2.25 ×, the less mobile crystals in the equatorial region of the spherulite are highly oriented ($f_c = 0.83$, see Fig. 3-24) and strongly resist further orientation. The small amount of remaining a-axis-oriented lamellae in the polar region are quickly drawn out, while the noncrystalline region supports the deforming stress by orienting at a rapid rate (slightly slowed as this region becomes very highly strained). By a draw ratio of 3.87 the crystals have not yet reached an f_c of 0.9 and thus seem to be resisting crystal-cleavage processes. This conclusion is supported by small-angle light scattering since an H_v SALS pattern was obtained at this high draw ratio (Figs. 3-28 and 3-29). The poor quality of the pattern (Fig. 3-29) is due to both the small fiber diameter and the high intraspherulite disruption. The highly strained noncrystalline chains have by now reached an f_{am} value of 0.7. Any further stress causes fracture under this processing condition, and hence 3.87 is the highest draw ratio attainable.

The highly oriented, strained, noncrystalline chains separate the resisting crystal lamellae and this is manifested by a gradual increase in the long spacing with extension in the high-draw region. Small-angle light-scattering patterns are obtained from all of the drawn fiber series in this draw region, showing that the spherulites have an identifiable boundary, but increasing distortion of the patterns with extension mirrors the internal disruption occurring in this highly strained system (Fig. 3-29). These structural conditions are schematically represented in Fig. 3-21 (high-draw region, drawing stage).

When these highly strained fibers are heat set under tension above the crystal mobility temperature, pronounced structural changes occur. The noncrystalline chains relax somewhat as the crystals become fully oriented (Fig. 3-24). The partial relaxation of the noncrystalline chains allows the long spacing to remain constant up to a draw ratio of 3.8 (Fig. 3-27). The increase in stress on the noncrystalline chains, in the highest draw ratio range of 3.8–4.48, causes an increasing orientation of these chains (Fig. 3-24) which, in turn, increases the long spacing in this upper draw region (Fig. 3-2) (see Chapter 4, Section A1.b for further discussion of the long-spacing data). When f_{am} reaches a value of 0.7, the maximum draw ratio under this processing condition is attained and any further drawing leads to fracture (Fig. 3-24).

The lamellae even at this high temperature are constrained by oriented noncrystalline chains. This results in crystal cleavage as the lamellae become fully oriented. The spherulitic character of the fiber substructure is lost and the structure becomes microfibrillar. The H_v SALS patterns characteristic of spherulite boundaries can no longer be found, and the SAXS patterns show layer line spreading (5). Heat setting under tension thus leads to a striking change from a structure of internally strained spherulites to one of microfibrillar character. These changes are schematically represented in Fig. 3-31.

3. Summary

A *modus operandi* has been developed for characterizing a multifilament polypropylene yarn process in quantitative morphological terms. This involves the application of eight different physical techniques. The changes that occur on the spherulitic, interlamellar, and molecular levels of fiber structure during the drawing of a spun fiber at 90°C (below the crystal-mobility temperature) were examined and compared with those observed for the same spun fiber after drawing at 90°C with subsequent heat setting under tension at 140°C (above the crystal-mobility temperature). These fibers, taken together, represent different stages of a single process: spinning, drawing, and heat setting. The morphological transformations occurring during the process were different in the high and low ranges of extension. These transformations can be summarized as follows.

1. In the low-draw region, during drawing, the crystal lamellae resisted orientation and the noncrystalline regions controlled the deformation. The spherulite deformation was not affine, major deformation occurring in the polar regions, as the *a*-axis-oriented crystals were drawn out. On heat setting, the most prominent feature was the crystal mobility imparted by the high temperature. This allowed the lamellae in the equatorial region of the spherulite to rotate into a more *c*-axis-oriented position, which caused the extension ratio of the spherulites to decrease slightly.

2. In the high-draw region, structural changes are largely controlled by the noncrystalline molecules. During drawing, the highly oriented crystals resist cleavage or further orientation, and the noncrystalline region orients to relieve the stress. The spherulites, though highly strained internally, keep their boundaries intact. On heat setting, the noncrystalline region relaxes slightly, allowing the lamellae to become fully oriented by crystal cleavage and slip processes. This leads to a transformation of the structure from spherulitic to microfibrillar.

References

1. R. J. Samuels, *J. Polymer Sci.*, *A-2*, **6**, 1101 (1968).
2. R. J. Samuels, *J. Polymer Sci.*, *A*, **3**, 1741 (1965).
3. S. Hoshino, J. Powers, D. G. LeGrand, H. Kawai, and R. S. Stein, *J. Polymer Sci.*, **58**, 185 (1962).
4. Z. W. Wilchinsky, *Polymer*, **5**, 271 (1964).
5. R. J. Samuels, in *Supramolecular Structure in Fibers* (*J. Polymer Sci. C*, **20**), P. H. Lindenmeyer, Ed., Interscience, New York, 1967, p. 253.
6. I. L. Hay and A. Keller, *Kolloid-Z.*, **204**, 43 (1965).
7. K. Kobayashi and T. Nagasawa, in *U.S.–Japan Seminar in Polymer Physics* (*J. Polymer Sci. C*, **15**), R. S. Stein and S. Onogi, Eds., Interscience, New York, 1966, p. 163.
8. W. A. Sisson, *Textile Res. J.*, **7**, 425 (1937).
9. P. H. Geil, in *Small Angle Scattering from Fibrous and Partially Ordered Systems:* (*J. Polymer Sci. C*, **13**), R. H. Marchessault, Ed., Interscience, New York, 1966, p. 149.
10. C. W. Hock, *J. Polymer Sci. B*, **3**, 573 (1965).
11. R. J. Samuels, *J. Polymer Sci.*, *A-2*, **9**, 2165 (1971).
12. R. J. Samuels, in *Small Angle Scattering from Fibrous and Partially Ordered Systems* (*J. Polymer Sci. C*, **13**), R. H. Marchessault, Ed., Interscience, New York, 1966, p. 37.

13. R. J. Samuels, *J. Polymer Sci.*, *A-2*, **6**, 2021 (1968).
14. H. D. Keith, F. J. Padden, Jr., and R. G. Vadimsky, *Science*, **150**, 1026 (1965).
15. H. D. Keith, F. J. Padden, Jr., and R. G. Vadimsky, *J. Polymer Sci. A-2*, **2**, 267 (1966).
16. R. J. Samuels, in *Science and Technology of Polymer Films*, O. J. Sweeting, Ed., Interscience, New York, 1968, Chap. 7.
17. M. Takayanagi, S. Minami, and H. Nagatoshi, *Asahi Garasii Kogyo Gijutsu Shoreikai*, **7**, 127 (1961).
18. J. A. Sauer, D. R. Morrow, and G. C. Richardson, *J. Appl. Phys.*, **36**, 3017 (1965).

4

Application: Quantitative Correlation of Polymer Structure with End-Use Properties

A. ISOTACTIC POLYPROPYLENE*

1. Mechanical Properties

a. Introduction. Isotactic polypropylene is a polycrystalline polymer. It has structural order on the molecular, interlamellar, and spherulitic levels. The character of this structure will vary with conditions of fabrication. Certainly the mechanical properties of a sample of isotactic polypropylene will depend on its particular structural arrangement.

One of the quantitative techniques developed for the purpose of characterizing molecular structure is sonic modulus. In this method (Chapter 2, Section B.3), a sound pulse propagates along the sample and its velocity is measured. The sound wave propagates by first compressing and then increasing the distance between molecules or between segments along a molecule, thus becoming a molecular stress–strain measurement of the modulus of the sample. This mechanical modulus measurement has been quantitatively correlated with the molecular structure in films of isotactic polypropylene (1,2), hydroxypropylcellulose (3), and poly(ethylene terephthalate) (4).

From these studies there is little doubt that at least one mechanical property of the polymer, its modulus, results as a direct consequence of the morphological structure of the sample and that a large amount of seemingly confusing modulus results can be easily understood in quantitative structural terms. There are other mechanical properties of polycrystalline polymers that are difficult to systemetize in purely mechanical terms. These include fracture strength (tensile strength of films, tenacity of fibers), tensile yield behavior, and tensile recovery (work and elastic recovery). The primary

* This section is reprinted from the *Journal of Macromolecular Science-Physics*, **B4**, No. 3, pp. 701–759 (1970), by courtesy of Marcel Dekker, Inc.

purpose of this section is to show how all of these properties are intimately involved with the structure of the sample being tested, and how large amounts of mechanical data can be quantitatively correlated through the use of structural criteria. Development of a relatively simple structural model makes it possible to see the direct relations between fabrication processes, mechanical tests, rate, and temperature, as well as between fibers and films.

b. A General Model for the System.   A cast film of isotactic polypropylene is spherulitic (Fig. 4-1). The spherulites are undeformed and have inter-spherulitic connections, which allow them to deform coherently under tension. Intraspherulitically, they are composed of crystalline (folded-chain lamellae) and noncrystalline regions. The crystal lamellae have a preferred orientation with respect to the spherulite radial direction and are connected by interlamellar ties (2,5).

When a stress σ is applied to such a film, the film will be deformed by a certain amount of strain ϵ. Under proper conditions of rate and temperature, a stress–strain curve will be obtained similar to that shown schematically in Fig. 4-2. The illustrations above the different regions of the stress–strain curve in the figure represent schematically the deformation processes occurring during drawing (2). Initially there is a sharp rise in load with increasing extension and the curve is approximately linear, corresponding to the initial modulus in the figure. In this region of very small extensions (0.5 % or less) the spherulites deform elastically as a unit. As stress increases the slope of the stress–strain curve usually falls slowly, that is, prior to yield, the curve is concave to the strain axis. This strain-softening region is the deformation region in which a small amount of nonrecoverable disruption of crystals occurs in the central region of the spherulite along radii parallel to the deformation direction; molecular tilting occurs within crystal lamellae in radii oriented at an angle to the deformation direction; and separation of lamellae occurs along radii in the equatorial region of the spherulite (Figs. 4-2 and 4-3) (2). Much of the deformation occurring in this region of the stress–strain curve is already plastic.

As stress builds up in the strained polymer the yield point is reached

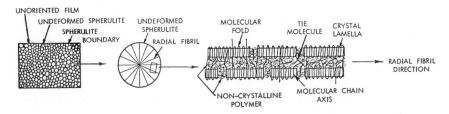

Fig. 4-1   Schematic representation of cast film structure.

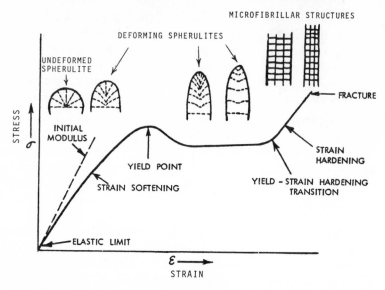

Fig. 4-2 Schematic representation of structural changes occurring in different regions of the stress–strain curve.

(Fig. 4-2). Following the yield point the stress decreases with increasing strain to some limiting value after which the stress remains constant with increasing strain over a fairly large region of strain. In the region of the stress–strain curve beyond the yield point the specimen under test is often found to have some local reduction in cross-sectional area—a phenomenon referred to as necking. The stress remains constant with increasing strain until the whole gauge length is reduced to a uniform cross section by this process. Structurally the yield point can be represented as the point at which shear forces within the spherulites predominate, leading to lamellar slip (Fig. 4-3). As strain continues beyond the yield point the spherulites are drawn out (Fig. 4-2) but preserve their continuity intraspherulitically. Within the spherulite several processes are occurring simultaneously. The amount of disruption increases along radii parallel to the deformation direction in the central region of the spherulite at the expense of the outer parallel-radii region. This is required in order to preserve continuity within the spherulite as the outer portion of the spherulite in the region intermediate to the deformation direction undergoes two types of deformation: (1) those off-axis radii subject to shear forces undergo lamellar slip and reorient with their long-spacing (chain-axis) direction parallel to the deformation direction, and (2) the distance between lamellae continues to increase with extension (2) (see Chapter 3, Section B).

The deformation of the spherulites continues by the processes of lamellar

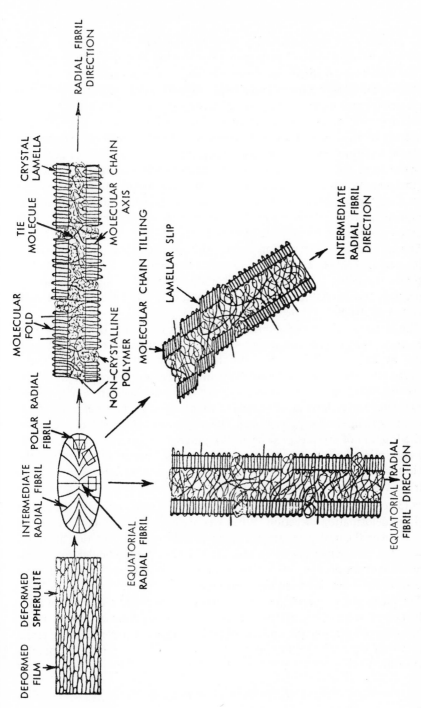

Fig. 4-3 Schematic representation of deformation processes in deformed spherulitic film.

slip, orientation, and separation until the lamellae in all regions of the spherulite became aligned with their long-spacing direction nearly parallel to the deformation direction. Once this stage of organization is reached, crystal orientation becomes exceedingly difficult, and further deformation must occur predominantly by crystal cleavage (2). This fundamental change in the deformation process results in a change in the stress–strain behavior. Once all of the spherulites have transformed to microfibrils subsequent increase in strain results in a continual increase in stress up to fracture of the specimen. This is represented by the strain-hardening region in Fig. 4-2. At the point of transition from yielding to strain hardening, the transition from a deformed spherulitic to a microfibrillar structure can be considered complete, hence the strain-hardening region of the stress–strain curve can be characterized as the fiberlike deformation region.

During this transformation from undeformed spherulites into first deformed spherulites and then a microfibrillar structure the noncrystalline region is also being deformed. It is the combination of both these deformation processes, the crystalline and the noncrystalline, that leads to the final anisotropic structural state of the polymer.

The stress–strain curve in Fig. 4-2 describes the limits of behavior of a specimen. That is, the unoriented state is at one extreme of the structural states available to the system and the microfibrillated state just before fracture is at the other. Different fabrication and test procedures can only move the specimen to various positions between these structural limits. Some structural criteria must be used to characterize the particular structural state of a specimen at any given time. The state of the system will be a function of the fraction of crystals present in the system β, the average orientation of the molecular chain axes in the crystal f_c, and the average orientation of the molecular chain axis of the molecules in the noncrystalline region f_{am} (1). Here f stands for the Hermans orientation function (Fig. 4-4), defined as $f \equiv (3 \cos^2 \theta - 1)/2$, where θ is the angle between the polymer-chain axis and the deformation direction. When the molecules are oriented randomly $f = 0$; when the molecules are oriented parallel to the deformation direction $f = +1.0$; and when the molecules are oriented perpendicular to the deformation direction $f = -0.5$. Other orientations of the molecules will have intermediate values between these extremes.

The measured properties of a specimen will be related either to the individual orientation of one of the components of the structure or to the average of both the orientation and amount of each component present. The average orientation state of the specimen f_{av} is defined here as (see Chapter 2, Section B.1):

$$f_{\text{av}} \equiv \beta f_c + (1 - \beta) f_{\text{am}} \tag{4-1}$$

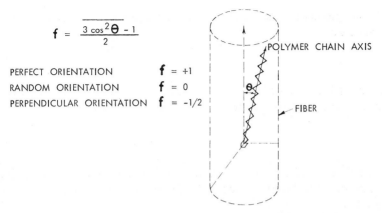

$$f = \frac{\overline{3 \cos^2 \Theta - 1}}{2}$$

PERFECT ORIENTATION	f = +1	
RANDOM ORIENTATION	f = 0	
PERPENDICULAR ORIENTATION	f = -1/2	

POLYMER CHAIN AXIS

FIBER

Fig. 4-4 Schematic representation of the significance of the Hermans' orientation function.

It is common to relate the properties of a specimen to the draw ratio the starting polymer underwent during specimen fabrication. However, the fabrication process, just as a tensile test, can be considered simply as a deformation process at a given rate and temperature which will deform the system to some state intermediate between the unoriented and the highly strained microfibrillar extremes. This means that different fabrication processes can lead to equivalent states of average orientation, and it is the orientation state of the polymer and not the fabrication draw ratio that should be used as a generalizing criterion.

The fabrication process, just as the stress–strain curve in Fig. 4-2, is primarily a plastic deformation process. It is preferable when considering plastic deformation to employ the expressions *true stress* and *true strain* instead of the *engineering stress* and *engineering strain* and *extension ratio* which are primarily applicable to plastic processes (6). Here *engineering stress* σ is defined as the ratio of the load on the sample P to the original cross-sectional area A_0:

$$\sigma = \frac{P}{A_0} \tag{4-2}$$

Engineering strain ϵ is defined as the ratio of the change in length of the sample Δl to its original length l_0:

$$\epsilon = \frac{l - l_0}{l_0} = \frac{\Delta l}{l_0} = (\lambda - 1) \tag{4-3}$$

where λ, the *extension ratio* (draw ratio), is the ratio of the final length l, to its original length ($\lambda = l/l_0$). The *true stress* σ_T is defined as the ratio of the load on the sample to the instantaneous minimum cross-sectional area A supporting that load:

$$\sigma_T = \frac{P}{A} = \sigma\lambda \tag{4-4}$$

The *true strain* ϵ_T is defined as the integral of the ratio of an incremental change in length to the instantaneous length of the sample:

$$\epsilon_T = \int_{l_0}^{l} \frac{dl}{l} = \ln\frac{l}{l_0} = \ln\lambda \tag{4-5}$$

For the plastic deformation processes involved in polycrystalline polymer deformation, it is the true strain (eq. 4-5) that must be related to structural parameters.

Samples from four different fabrication processes are examined here [These are the Series A and Series B films (see Chapter 3, section B) and the Series D (drawn) and Series E (heat set) fibers (see Chapter 3, section C)], and it is interesting to see how the average orientation, the crystal orientation, and the noncrystalline orientation are related to their respective fabrication draw ratios. In Figs. 4-5 to 4-7 are shown f_{av}, f_c, and f_{am}, respectively, plotted

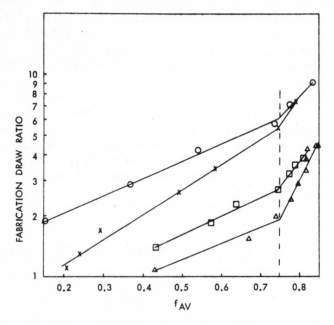

Fig. 4-5 Relation between fabrication draw ratio and the average orientation in isotactic polypropylene fibers anh films: (O) film (draw temperature 135°C); (×) film (draw temperature 110°C); (□) fiber (draw temperature 90°C); (△) fiber (heat set).

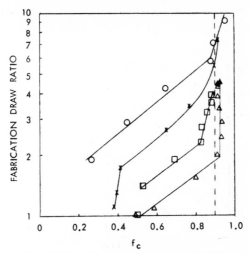

Fig. 4-6 Relation between fabrication draw ratio and the crystal orientation in isotactic polypropylene fibers and films: (○) film (draw temperature 135°C); (×) film (draw temperature 110°C); (□) fiber (draw temperature 90°C); (△) fiber (heat Iet).

against the fabrication draw ratio (λ_{fab}) in the form of a true strain. Several features of these curves are noteworthy. First, in all cases except one (the f_c curve for film drawn at 110°C) the orientation is a linear function of the true fabrication strain. The noncrystalline orientation function f_{am} is essentially linear with no change in slope over the whole draw range for all of the fibers and films. The crystal-orientation function (with the one exception) is linear up to a value of approximately 0.9. A break then occurs with a new line of high slope—this is the region of the spherulitic-to-microfibrillar transition (2,7,8). The average orientation function, which also considers the fraction of each component present in the sample, can easily be seen as the weighted sum of the curves in Figs. 4-6 and 4-7. The average orientation for all fabrication processes is linear with true fabrication strain, with the point of intersection of the two linear portions of each curve at a value of 0.76. This value of f_{av} represents the boundary orientation between the region where the specimen has spherulitic character and that in which a microfibrillar structure predominates.

One assumption of the structural model, namely, that different fabrication processes can lead to equivalent states of average orientation, is immediately apparent from these figures. Thus, from Fig. 4-5, the fabrication draw ratio at which the transition from a spherulitic to a microfibrillar structure occurs can have values ranging from 1.9 to 6, depending on the temperature and conditions of drawing. Similarly, other equivalent states of crystalline and noncrystalline orientation are produced at different values of the fabrication

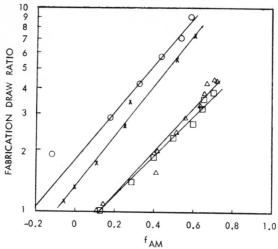

Fig. 4-7 Relation between the fabrication draw ratio and the amorphous orientation in isotactic polypropylene fibers and films: (○) film (draw temperature 135°C); (×) film (draw temperature 110°C); (□) fiber (draw temperature 90°C); (△) fiber (heat set).

draw ratio depending on the fabrication conditions utilized in preparing the different fiber and film samples. It is this equivalency of orientation states of samples produced under vastly different fabrication conditions that will make it possible to generalize a large amount of physical data in a systematic fashion.

An examination of the influence of fabrication conditions on the average repeat distance of the crystal lamellae [$L(\text{Å})$], as measured parallel to the helical chain axis of the molecules, will help to illustrate the importance of orientation parameters. Figure 4-8 shows the average long period of the crystal lamellae as a function of the fabrication draw ratio for the several fabrication processes used to prepare the fibers and films. Each curve in the figure is obviously unique, and there is little to suggest any generalizing correlations between the different systems. The same long-spacing data are plotted in Fig. 4-9 as a function of the average orientation in the sample. Correlation between the behavior of the different samples is immediately obvious. Thus, a major transition in behavior of all of the fibers and films occurs at an f_{av} value of 0.76, which is the transition region between a spherulitic and a microfibrillar texture. Beyond an f_{av} value of 0.76, each of the curves level off. The value of $L(\text{Å})$ in the region beyond $f_{av} = 0.76$ is seen to be some function of the draw temperature (the heat-set fiber was drawn at 90°C). These long spacing data are superimposed in Fig. 4-10 on a curve of long spacing versus draw temperature recently obtained by Balta-Calleja and Peterlin (9) from a sample of cast polypropylene film. These authors

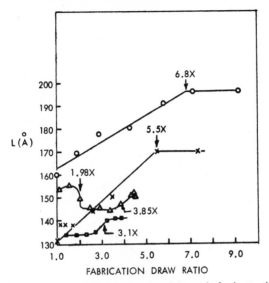

Fig. 4-8 Long spacing as a function of fabrication draw ratio for isotactic polypropylene fibers and films: (○) film (draw temperature 135°C); (×) film (draw temperature 110°C); (□) fiber (draw temperature 90°C); (△) fiber (heat set).

found that if they stretched their film sample at a rate of 10%/min, to an extension ratio of 5 or greater (in order to produce a microfibrillar texture), at the different temperatures, they would always obtain a single long spacing at a given temperature, irrespective of the long spacing in the starting film. The curve in Fig. 4-10 contains the long-spacing draw temperature behavior that they observed. The long-spacing data obtained from Fig. 4-9 fit exactly

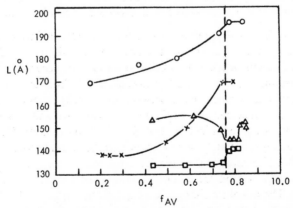

Fig. 4-9 Long spacing as a function of average orientation in isotactic polypropylene fibers and films: (○) film (draw temperature 135°C); (×) film (draw temperature 110°C); (□) fiber (draw temperature 90°C); (△) fiber (heat set).

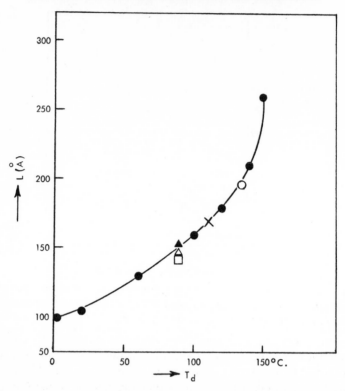

Fig. 4-10 Long spacing as a function of draw temperature: (O) film (draw temperature 135°C) $f_{av} \geqslant 0.76$; ($\times$) film (draw temperature 110°C) $f_{av} \geqslant 0.76$; ($\square$) fiber (draw temperature 90°C) $f_{av} \geqslant 0.76$; ($\triangle$) fiber (heat set) $f_{av} = 0.76$; ($\blacktriangle$) fiber (heat set) $f_{av} = 0.83$; ($\bullet$) cast film, drawn $\lambda \geqslant 5$ at temperatures shown [Peterlin (9)].

on the line in Fig. 4-10, again demonstrating the equivalence of a fabrication process to a typical stress–strain measurement. The figure also demonstrates that the equivalent process of microfibrillation can occur at different draw ratios depending on the method of preparation. These data suggest that a single general relation may exist between other properties and temperature for all polypropylene samples in the microfibrillar structural state.

Two points are shown in Fig. 4-10 for the heat-set fiber. One represents the long spacing value at $f_{av} = 0.76$, whereas the second represents the value at $f_{av} = 0.83$. The value at 0.83 results in the sample as a consequence of the high strain imposed on the noncrystalline chains under the conditions of fabrication. The absolute change in long spacing on the scale of Fig. 4-10 is small, however. Another interesting feature of the fiber curves in Fig. 4-9 is that before a value of $f_{av} = 0.76$ the long-spacing values are lower than those after 0.76 for the drawn-fiber samples, but higher than those after

0.76 for the heat-set samples. The fact that there is a thermodynamically stable long spacing for a given draw temperature in the microfibrillar state is dramatically demonstrated here, as both systems converge to a similar long spacing at $f_{av} = 0.76$.

The important features of the foregoing model are that (1) there are many different fabrication paths leading to equivalent average orientation states; (2) all average orientation states will fall somewhere between the unoriented spherulitic state and the highest oriented microfibrillar state attainable before flaw mechanisms cause material failure; (3) the plastic deformation characteristics of polycrystalline polymer deformation require that true strain and not draw ratio be correlated with the average orientation state; and (4) the state of orientation and not the fabrication draw ratio should be used to characterize mechanical properties of fibers and films.

In this section a mechanical property will be related to the structural state of the material. It is, therefore, important that the property be defined in structurally meaningful terms. For example, the accepted measure of the strength of a material is its tensile strength. The tensile strength is conventionally defined as the load on a sample at the point on the stress–strain curve where the stress reaches a maximum, divided by the original cross-sectional area of the sample. As can be seen from Fig. 4-11a, if a sample is deformed at a rate and temperature that cause failure before the material can strain harden, this tensile strength will be the strength at the yield point divided by the original cross-sectional area of the specimen. On the other

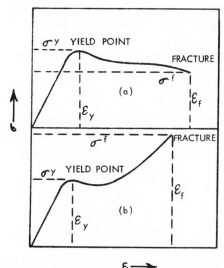

Fig. 4-11 Stress–strain behavior of ductile polymers (a) without and (b) with strain hardening.

hand, if strain hardening does occur (Fig. 4-11b), then the tensile strength by the standard definition will be the force at the point of fracture divided by the original cross-sectional area of the specimen. Calling both these values the tensile strength will be misleading in any attempt at structural correlations. In this study the tensile strength σ^f will always be the force at the point of fracture divided by the original cross-sectional area of the sample.

For fracture studies the strength of the material is desired in terms of the structural state of the sample at the time of fracture (i.e., the strength of the material in terms of the cross-sectional area of the specimen at the time of failure). This important parameter is the *true stress at failure*, σ_T^f, as defined by eq. 4-4. The σ in eq. 4-4 is now σ^f, as defined above, and λ is the extension ratio at the point of fracture, $\lambda_f (\sigma_T^f = \sigma^f \lambda_f)$. The yield stress σ^y is defined in this study as the tensile force at the yield point divided by the original cross-sectional area of the specimen (Fig. 4-11). With these definitions it is now possible to consider the relations existing between the mechanical properties of uniaxially oriented isotactic polypropylene fibers and films and the internal structural state of the material.

2. Failure Mechanics

a. Structural Theory. In any sample of a polymer it is assumed that there is a distribution of flaws (stress concentration regions) that, under the proper deformation conditions, can cause crack formation and ultimate failure. Opposed to the failure mechanism is the ability of the sample to deform plastically to relieve and redistribute the stress. High temperature and slow rate of deformation allow the molecular motions necessary for plastic deformation. Low temperature and high rate of deformation do not allow time for the molecules to rearrange, and the flaw mechanism of failure predominates.

If a sample is deformed slowly enough and the temperature is high enough, a stress–strain curve similar to that shown schematically in Fig. 4-2 can be obtained. The molecules have enough mobility and time available for rearrangement of the internal structure to relieve the stress. The spherulitic structure is slowly destroyed, a microfibrillar structure develops, stress builds up in the microfibrillar structure as the noncrystalline chains get more oriented, and finally the flaw mechanism predominates as no further plastic deformation mechanisms are available and the sample fractures. If the process of deformation is slow enough to allow the necessary time for all of these processes to occur, the strength of the sample at the moment of failure will be the same, independent of its starting state of orientation. That is to say, the final structure at break will be the microfibrillar structure thermodynamically stable at the draw temperature at the moment the failure mechanism takes control, and that final structure will be independent of the starting structure. Thus the strength of the sample, with respect to the final cross-sectional area at break (i.e., the true stress) will be independent of the initial orientation state of the sample.

On the other hand, if a sample is drawn very rapidly and at low temperature, so that no motion is available to the molecules, the flaw mechanism will predominate, the sample will only show extension due to crack formation, and the strength of the specimen will be solely a function of the support contributed by those polymer chains already oriented in the direction of the applied stress. The greater the number of chains oriented in the direction of the applied stress, the greater will be the force required to fracture the sample, that is, the strength will be a direct function of the state of orientation of the specimen before deformation.

Between these two extremes there is a region of transitional behavior wherein some plastic deformation can occur before the failure mechanism takes control. Behavior in this intermediate region will be a function of the rate of deformation, the temperature of deformation, and the initial state of orientation of the system. The importance of the initial orientation state can be seen from an examination of the effect of orientation on molecular mobility as observed from nuclear magnetic resonance, dynamic mechanical, and thermal distortion studies on crystallizable polystyrene (10), nylon 66 (11–13), and poly(ethylene terephthalate) (14,15). These studies showed that there is an increase in resistance to thermal mobility of the molecules with increasing orientation, and the higher the molecular orientation, the greater the observed glass transition temperature T_g. Thus, at a given temperature of measurement, a highly oriented sample will act as if its environment was at a lower temperature (the molecules are stiffer) than that of an unoriented sample measured at the same temperature.

If the above model is correct, it should be possible to obtain a fracture envelope for isotactic polypropylene. At one extreme of the envelope will be the slow rate process where fracture strength is independent of the initial orientation state of the polymer, whereas the fast rate processes that are orientation dependent will be found at the other extreme. The two regions should meet at the highest average orientation where extension is no longer available as a mode of response. Between these extremes would be a transition region that is rate dependent. This hypothesis is examined experimentally in the next section.

b. The Fracture Envelope.

(1) Rate and Initial Structure Dependence.   Figure 4-12 shows the true stress fracture envelope for the Series A film. Five specimens of each orientation state were elongated at room temperature at the rates of strain listed in the figure. Only two orientation states were tested at a rate of 1 %/min since

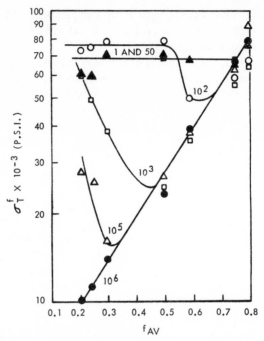

Fig. 4-12　True stress fracture envelope for Series A isotactic polypropylene film (5 samples each point).

it takes approximately 10 hr at this temperature and rate to obtain failure from a single specimen. The fracture envelope is plotted in the same semilogarithmic form required for structure correlations with extension.

The general features of the experimental true stress fracture envelope are in agreement with the predictions of the structural theory. At low rates of deformation (1 and 50%/min), σ_T^f is independent of the initial state of orientation of the sample (f_{av}). At the highest rate of extension ($10^6\%$/min), σ_T^f is a direct function of the initial state of orientation of the sample. These envelope boundaries intersect at high average orientation, again as predicted, and there is a rate-dependent transition region between the boundary limits.

The character of the changes in the rate-sensitive transition region follows expectations. The more oriented the system, the stiffer will be the molecules and, hence, the more resistant they will be to deformation. As such, the more oriented molecules will respond to a given rate of deformation as though they were being pulled at a faster rate than molecules deformed at the same rate in an unoriented specimen. This means the more highly oriented the specimen, the greater its rate sensitivity. Thus, in Fig. 4-12, at a draw rate of $10^2\%$/min, the first samples that undergo a transition to the high-rate boundary line are the most oriented samples. As the rate of deformation increases, lower orientation samples undergo a transition in their mode of fracture until, at a rate of $10^6 \%/\text{min}$, all of the samples fracture as a direct function of their initial orientation, that is, $\log (\sigma_T^f \sigma_{T,0}^f) = A f_{\text{av}}$.

A true strain fracture envelope is also predicted by the structural theory. When the samples are deformed slowly, so they can all reach the same boundary microfibrillar state, they will each have a definite extension to which they must deform and that extension is determined by their initial state of orientation. The unoriented film will have the highest extension ratio, for it must travel the total length of the stress–strain curve (Fig. 4-2) before fracture. A microfibrillar sample will have little extension since it is already in a strain-hardened structural state. Other structural states will have intermediate extensions between these extremes.

Similarly, at high deformation rates the molecules do not have much mobility and, hence, all the samples will manifest little extension. Between these rate extremes will be a transition strain region characterizing the interplay between the plastic deformation and fracture mechanisms. Thus, the characteristics of the true strain fracture envelope will be the inverse of those of the true stress fracture envelope.

Figure 4-13 shows the true strain fracture envelope for the Series A film. As expected this envelope looks like the inverse of the true stress fracture envelope (Fig. 4-12). The true strain ($\ln \lambda_f$) at slow rates of extension (1 and $50\%/\text{min}$) is proportional to the initial orientation state of the sample, whereas at a high rate of extension ($10^6 \%/\text{min}$) the true strain is independent of the initial sample orientation state. These two boundary curves meet at a high state of initial average orientation, and there is a transition region between the two boundary curves. Also the rate sensitivity of the true strain in the transition region to the average initial orientation in the sample follows a similar pattern to that of the true stress in the true stress fracture envelope.

The agreement, in Figs. 4-12 and 4-13, of the experimental fracture envelopes with the predictions from the structural deformation model suggested that the acquisition of more quantitative data was in order. In particular it seemed pertinent to measure the failure of a large number of samples (80) for each state of orientation. This would not only lead to more reliable average values of the true stress and true strain but would also result in representative distributions of fracture for each orientation state under the several test conditions.

The true stress fracture envelope and true strain fracture envelope for the Series A films, stretched at room temperature at the rates of deformation listed, are shown in Figs. 4-14 and 4-15, respectively. Each point in the figures is an average of the measurement of 80 samples. In general character, these envelopes are the same as those obtained using five samples for each point (Figs. 4-12 and 4-13). The stress–strain values at rates of deformation of 1 and 50%/min were not determined because of the prohibitive time involved in measuring eighty samples of each orientation state at these rates.

(2) *Temperature Dependence.* The temperature of deformation is another parameter in the fracture process. In the above study the rate of deformation

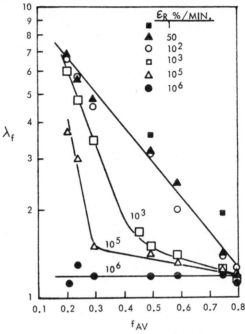

Fig. 4-13 True strain fracture envelope for Series A isotactic polypropylene film (5 samples each point).

was used to change the characteristic fracture behavior of the sample by moving the sample faster than molecular motion could follow. In this way those samples that, at 23°C, were above the glass transition temperature (i.e., their noncrystalline chains were mobile) could be made to act as if they were stiffer by an increase in the rate of deformation. Lowering the temperature is another technique used to decrease the mobility of the molecules (making them stiffer). It should be possible, if time and temperature are equivalent, to find a low temperature at a rate of 10^2%/min, at which the system behaves the same during failure as it does at 23°C at a rate of 10^6%/min. Since decreasing the temperature does increase the stiffness of the molecules, this is not an unreasonable assumption. The time-temperature superposition principle has been widely successful in the treatment of amorphous polymers

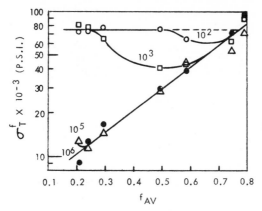

Fig. 4-14 True stress fracture envelope for Series A isotactic polypropylene film (80 samples each point).

(16). This does not necessarily mean, however, that the same system will be valid for crystalline polymers, where there are changes in structure and in the stress-bearing mechanisms with temperature and where the overall structural conditions do not conform to the theoretical model for viscoelastic behavior of amorphous materials.

The true stress and true strain fracture envelopes for the Series A samples, respectively, are presented in Figs. 4-16 and 4-17 as a function of rate and temperature. Two curves are shown for 23°C deformation—the curve obtained at a rate of deformation of 10^2 %/min and the curve obtained at a rate of deformation of 10^6 %/min. At a rate of 10^2 %/min the rate of deformation is slow enough at 23°C to allow structural rearrangements into entirely new microfibrillar structures before fracture. At 10^6 %/min the deformation at 23°C is so fast that little extension is available to the molecules and the true

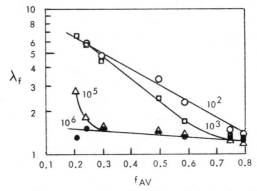

Fig. 4-15 True strain fracture envelope for Series A isotactic polypropylene film (80 samples each point).

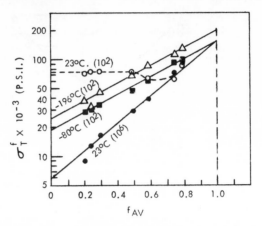

Fig. 4-16 Effect of temperature on the fracture behavior of the Series A isotactic polypropylene film (80 samples per point).

stress is proportional to the initial orientation in the sample. These two rates, in fact, encompass the boundaries of the fracture envelopes in Figs. 4-14 and 4-15.

When the temperature of deformation is decreased to $-80°C$ at a deformation rate of 10^2 %/min the samples become stiffer and their true stress at fracture decreases from what it had been at the same rate at room temperature (Fig. 4-16). However, even at this low temperature there is a significant amount of extension in the less oriented samples [almost 200% extension at an f_{av} of 0.207 (Fig. 4-17)] and the trend in true strain with initial orientation is similar to that observed at room temperature at a rate of 10^5 %/min.

The true stress at $-80°C$ is directly proportional to the average initial orientation of the sample, as it is at room temperature at a rate of deformation of 10^6 %/min, but the absolute values of the true stress are considerably higher than those observed at room temperature at 10^6 %/min. The observed increase in true stress at a lower temperature over that obtained at room temperature at 10^6 %/min is not due primarily to the observed higher strains at this temperature but, instead, is due to the greater strength of the chains at lower temperature (i.e., increased stiffness). That is, the decrease in thermal motion due to the decrease in temperature makes movement of the molecules more difficult at $-80°C$ at a rate of 10^2 %/min than at a high rate of deformation (10^6 %/min) at room temperature.

This effect becomes even more obvious when the temperature is lowered to $-196°C$ (77°K). The decrease of 116°C in going from a temperature of $-80°C$ to $-196°C$ has had little effect on the extension of the samples as a function of their initial orientation (Fig. 4-17). There is a slight drop in true strain, which, if the strain was the predominant contributor to the increase in true stress, would lead to a decrease in true stress from that observed at −80°C. In fact, there is an increase in true stress at −196°C over that observed at −80°C. Thus the primary effect of lowering the temperature is to increase the strength of the chains. It is interesting to note that the sample with the lowest initial orientation ($f_{av} = 0.207$) has a high elongation (136%) at −196°C and that the true stress of the samples at −196°C (Fig. 4-16) is a direct function of their initial average orientation.

c. Failure Distributions.

(1) Rate and Initial Structure Dependence. The primary purpose of measuring the failure of a large number of samples (80) for each orientation state was to examine the distribution of breaks as a function of the true stress at fracture. In considering the statistics of fracture of real materials, it is assumed that the scatter of results observed in actual physical testing is a consequence of the presence of a distribution of flaws (defects) in the sample. According to weak-link theory, failure will occur in each specimen at the location of the most serious flaw. In real samples these flaws are rarely visible and their number, relative size, and distribution are all unknown. Only their effect is observed as a scatter of test results (17).

The distribution of breaking strengths for the different orientation states of the Series A films during room-temperature extension are plotted at the respective draw rates in Figs. 4-18 to 4-21. The breaking-strength distribution

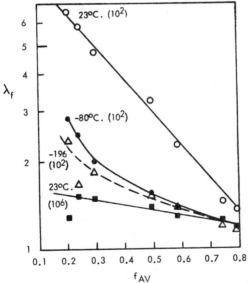

Fig. 4-17   Effect of temperature on the true strain envelope of the Series A isotactic polypropylene film (80 samples per point).

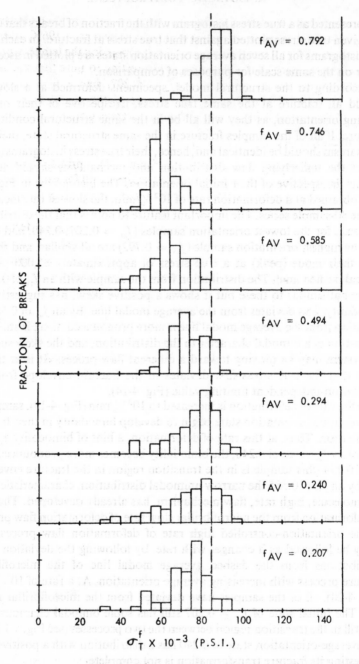

Fig. 4-18 True stress histograms for Series A films at a strain rate of $10^2 \, \%/min$ at 23°C.

is represented as a true stress histogram with the fraction of breaks that occurs at a given true stress plotted against that true stress at fracture. In each figure the histograms for all seven average orientation states are plotted in ascending order on the same scale for purposes of comparison.

According to the structural model, specimens deformed at a slow rate should all fracture at the same true stress, irrespective of their original starting orientation, as they will all be in the same structural condition at fracture. If all of the samples fracture in the same structural state, their flaw mechanism should be identical and, hence, their true stress histograms (which reflect the underlying flaw distribution and mechanism) should also be similar, irrespective of their initial orientation. The histograms in Fig. 4-18 were obtained at a deformation rate of 10^2 %/min, the slowest rate measured for the 80-sample series. The important feature to note is that the distribution of breaks for the lowest orientation samples ($f_{av} = 0.207, 0.240$, and 0.294) and the highest orientation sample ($f_{av} = 0.792$) are all similar, and they all have their mode (peak) at a true stress of approximately 85,000 psi (see vertical dashed line). The distribution from the sample with an f_{av} of 0.494 is somewhat similar to these but it shows a positive skew, has suggestions of bimodality, and deviates from the average modal line. By an f_{av} of 0.585 the deviation from the average modal line is more pronounced, there is still some suggestion of a bimodal character to the distribution, and the skew suggests the system may be moving toward a different flaw process. It is at this f_{av} value that the true stress first deviates in the fracture envelope from the orientation-independent limiting value (Fig. 4-14).

As the rate of deformation is increased to 10^3 %/min (Fig. 4-19), samples in a lower initial orientation state begin to develop bimodality in their fracture distribution. Thus, at this rate of deformation, a hint of bimodality appears at the low f_{av} value of 0.240, the bimodality becomes more pronounced at an f_{av} of 0.294 (this sample is in the transition region in the fracture envelope), and by an f_{av} of 0.585 the narrow unimodal distribution, characteristic of the stiff molecule, high rate, flaw mechanism, has already developed. Thus, the transformation from the microfibrillar slow rate of deformation flaw process, to the orientation-controlled high rate of deformation flaw process, can easily be followed as it changes with rate, by following the deviation of the distributions from the dashed average modal line of the microfibrillar fracture process with increasing average orientation. At a rate of 10^5 %/min (Fig. 4-20), all of the samples have deviated from the microfibrillar modal line. The initial state of $f_{av} = 0.207$ still has some bimodal character since it is still in the transition region between the two processes (see Fig. 4-14), and the average-orientation state of 0.240 has a distribution with a positive skew suggesting its fracture transformation is not complete.

By a deformation rate of 10^6 %/min at room temperature (Fig. 4-21), the

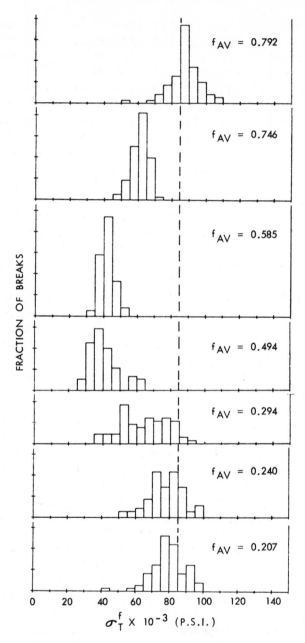

Fig. 4-19 True stress histograms for Series A films at a strain rate of 10^3 %/min at 23°C.

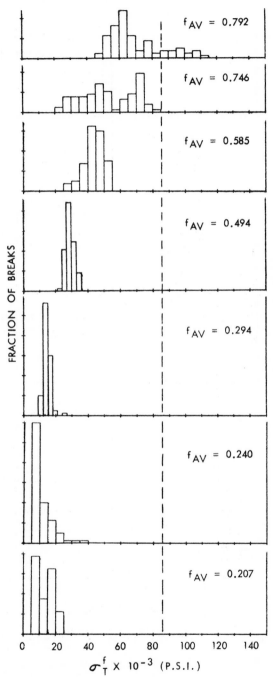

Fig. 4-20 True stress histograms for Series A films at a strain rate of 10^5 %/min at 23°C.

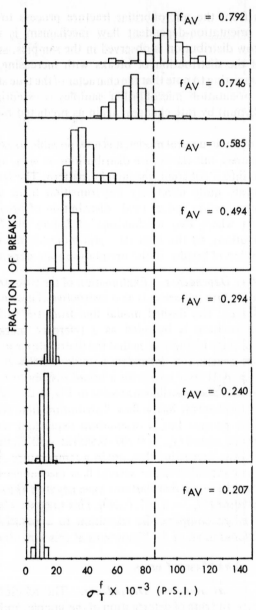

Fig. 4-21 True stress histograms for Series A films at a strain rate of $10^6\ \%/\text{min}$ at 23°C.

transformation from the microfibrillar fracture process to the high-rate, stiff-molecule, orientation-dependent flaw mechanism is complete. The unimodal, narrow distribution is observed in the samples, and the trend in the position of the distribution of breaks with increasing orientation is obvious. It is interesting to note that the character of the true stress histogram for the higher orientation microfibrillar samples is essentially unaffected (except at $10^5\%$/min) by rate changes, again as predicted by the structural model.

Thus, by use of the structural model, it is now possible to examine not only the failure envelopes, but the failure distributions as well, and to organize the information into a coherent, meaningful pattern. The true stress histograms demonstrate quite effectively the transition from a slow rate of deformation microfibrillar failure (wide distribution of breaks), through a transition region where two mechanisms of failure vie for dominance (bimodal distribution), to the high-rate, stiff-molecule failure mechanism (narrow distribution of breaks, initial orientation state-dependent).

(2) *Temperature Dependence.*   Examination of the true stress histograms for the low-temperature processes is also instructive. These are presented in Figs. 4-22 and 4-23. The dashed modal line from the $10^2\%$/min, room temperature deformation is included as a reference marker. The most striking feature of these histograms is that the distributions at both -80 and $-196°C$ have an orientation dependence similar to the $10^6\%$/min, $23°C$ histograms in Fig. 4-21, but now with a broad distribution of breaks like those found with the microfibrillar fractures in Fig. 4-18. This suggests the underlying flaw mechanism has a flaw distribution that is more like the microfibrillar flaw process but is orientation dependent. In the oriented spherulitic structural region ($f_{av} = 0.207$–0.585) at $-80°C$, the distributions have a positive skew, suggesting they are in a temperature region in which a transition in flaw mechanisms, or several flaw mechanisms, is occurring. By $-196°C$, all of the break distributions have negative skew except for the two lowest orientations ($f_{av} = 0.207, 0.240$). This suggests a further decrease in temperature might complete the transition to a negative skew of the distribution of breaks in all of the histograms at this slow-draw rate.

d. True Stress Failure Master Curves.

(1) *Rate and Initial Structure Dependence.*   The relation between the true stress at break, the rate of deformation of the sample, and the molecular orientation in the sample can be further examined as a viscoelastic problem using reduced variables. This general approach has been very successful in the study of viscoelastic properties of amorphous polymers as a function of rate (frequency) and temperature (16). A log–log plot is constructed of the

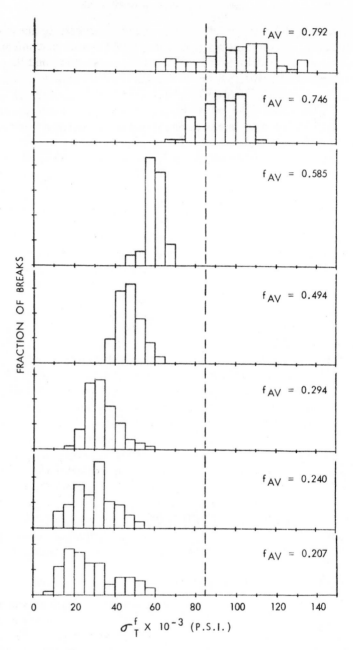

Fig. 4-22 True stress histograms for the Series A films at a strain rate of 10^2 %/min at $-80°C$.

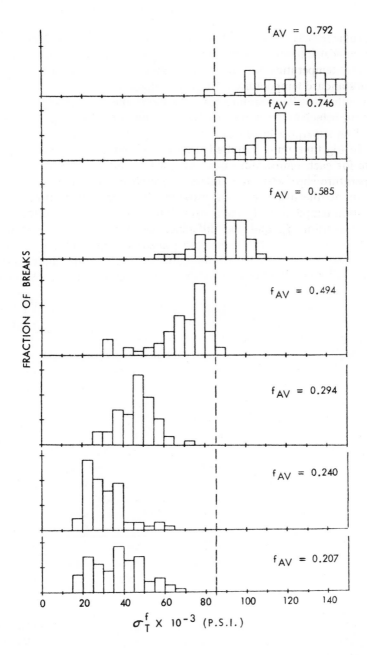

Fig. 4-23 True stress histograms for the Series A films at a strain rate of 10^2 %/min at $-196°C$.

187

particular property (stress to break, modulus, etc.) as a function of frequency, for a range of different temperatures. One of the temperatures is chosen as the standard temperature and the property parameter is multiplied by the ratio of the standard temperature to the test temperature to reduce all of the data to an equivalent temperature scale. Invoking the time–temperature super-position principle the curves are then slid horizontally along the time axis until they superpose positions to form a single master curve. The amount of slip, in log time units, required to move the reduced property–frequency curve for each temperature onto the master curve is also recorded. This temperature slip factor a_T is related through the Williams–Landel–Ferry relation to the distance of the experimental temperature from the glass transition temperature T_g, the fractional free volume of the molecules at the glass transition f_g, and the difference between the thermal expansion coefficients above and below the glass transition temperature a_2. In this way it is possible to obtain experimental information about the major transitions and relaxations of the amorphous chains in a reasonable time scale and relate these to molecular processes.

The study of the dynamic mechanical response of crystalline polymers has been partly successful in the very low deformation region (18), although transition and relaxation assignments are often speculative. Though failure envelopes have been obtained from amorphous polymers (19), none have been possible with crystalline polymers. The structural model in Sections A.1.b and A.2.a along with the quantitative structural criteria now available resolved this problem, and the resulting failure envelopes are discussed above. Knowing quantitatively the initial state of the material also makes it possible to develop a structure-rate master failure curve for isotactic polypropylene.

The general structural model developed in this study considers that there are a limited number of average-orientation states available to the crystalline polymer, that is, f_{av} can range from a value of zero to a value of $+1.0$. If the fully oriented state is defined as the limiting state, then the distance of any average orientation state f_{av} from the limiting orientation state f_{av}^L is simply $(f_{av}^L - f_{av})$. Since f_{av}^L is by definition equal to unity, this orientation reduction term (i.e., it defines the state of the material in terms of the ideal state) is really $(1 - f_{av})$. Figure 4-24 shows a log–log plot of the true stress at break σ_t^f multiplied by the structural-state reduction term $(1 - f_{av})$, plotted against the strain rate of room temperature deformation of the sample ϵ_r [here $(100\epsilon_r)$ simply means percent strain].

By holding fixed the position of the reduced true stress–strain rate curve of the sample having the lowest available orientation state ($f_{av} = 0.207$) and by slipping the other orientation-state curves along the time axis, it is possible to obtain the true stress failure master curve shown in Fig. 4-25. Here a_f is the structure slip factor each curve was moved along the log reciprocal time $(1/t)$ axis. Since strain rate is in reciprocal time, the smaller the number on the time axis the longer the time available for molecular movement.

The master curve is composed of three distinct regions. At slow rates of deformation when long times are available for molecular motion, the structure rearranges to a microfibrillar structure before failure. This corresponds to the upper limit line in the true stress failure envelope (Fig. 4-14). As the rate of deformation is increased, or the initial orientation of the material at a given rate increases, a region is reached where the molecules cannot move fast enough to keep up with the deformation. This is the transition region in the master curve and corresponds to the transition region in the true stress failure envelope (Fig. 4-14). The transition region also corresponds to regions of bimodality in the true stress histograms (Figs. 4-18 to 4-21). In a time–temperature superposed master curve, this transition region would be equivalent to the glass transition region. As the rate of deformation or the initial orientation state increases further, the molecules have little available motion and a third region of behavior appears in which the strength of the sample depends on the number of chains available for supporting the load. Thus, the strength in this region is orientation dependent. The master curve in the high-rate, high-orientation region is essentially flat for this reason, as the increase in strength due to the increased initial orientation is exactly balanced by the orientation state factor $(1 - f_{av})$. The behavior of the samples

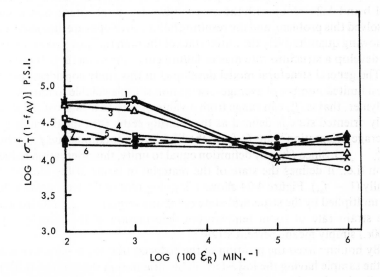

Fig. 4-24 Structure factor reduced true stress versus strain rate curves for the Series A films at room temperature. f_{av} for the representative numbers are: (1) 0.207, (2) 0.240, (3) 0.294, (4) 0.494, (5) 0.585, (6) 0.746, (7) 0.792.

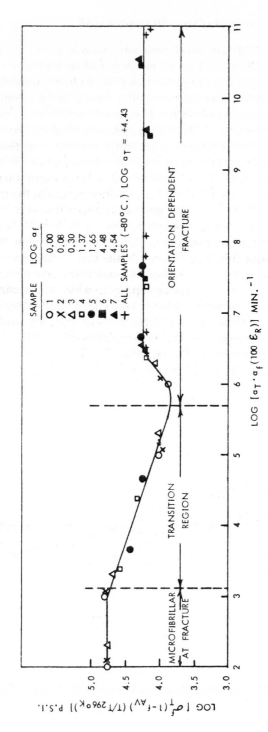

Fig. 4-25 True stress failure master curve for the Series A films.

in the third region of the master curve corresponds to the lower-limit, high-rate line in the true stress fracture envelope in Fig. 4-14.

Since the master curve is a structure–time plot, the orientation slip factor a_f should be some function of the orientation state of the material, just as the temperature slip factor a_t is related to the glass transition temperature of amorphous polymers. Log a_f is plotted against f_{av} in Fig. 4-26. The orientation slip factor is seen to be structure dependent. For those samples that were initially spherulitic in character, log a_f is a linear function of f_{av}. For samples having a microfibrillar structure the reduced true stress is independent of the rate of deformation (horizontal line in the true stress failure master curve) and the orientation slip factor can lie anywhere beyond a log a_f value of 4.477 min^{-1}. That is, the minimum value of log a_f for the microfibrillar samples is 4.477 min^{-1}. A log a_f value of 4.544 min^{-1} was chosen for $f_{av} = 0.794$ for fitting purposes; however, this is almost identical to the minimum microfibrillar a_f value. The minimum microfibrillar a_f value is outside the linear region of the log $a_f - f_{av}$ curve characteristic of an initial spherulitic structure and suggests a different structure–slip factor relationship must be applied. This result is consistent with previously observed differences in the structural characteristics of a spherulitic as compared to a microfibrillar structure (see, for example, Figs. 4-5 and 4-9).

(2) *Temperature Dependence.* It is interesting to note that all of the

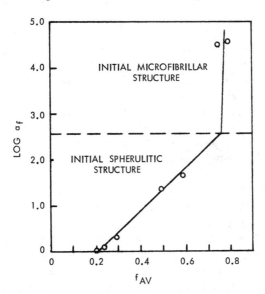

Fig. 4-26 Relation between the orientation slip factor a_f and the average initial orientation f_{av} of the Series A films.

samples having a spherulitic character undergo yielding at the low temperatures of $-80°$ and $-196°C$ before they fracture (see Section A3). The samples
must, therefore, be considered as ductile, as is also attested to by the high
fracture strains. This suggests considerable mobility is available to the
molecules even at -80 and $-196°C$. Dynamic mechanical and nuclear-
magnetic-resonance studies of transition and relaxation phenomena in
polypropylene support this view. The mechanical damping results obtained
by Turley and Keskkula from isotactic polypropylene (20) using a torsion
pendulum with a frequency of oscillation of 1 Hz will serve as an example. A
change in frequency would displace the observed transition temperature
values but the general temperature region would be the same. Turley's
logarithmic decrement curve shows the presence of three maxima. One is a
high-temperature ($+60°C$) crystal-disordering transition. A second maximum occurs at $5°C$ and is associated with the glass transition of polypropylene. The room-temperature deformations of the Series A film were
carried out at a temperature between these two transition temperatures.
Thus, long-chain segments (~ 100 carbon atoms) of molecules in the noncrystalline region had considerable mobility, and slow rates of deformation
at this temperature enhanced crystal mobility and ductility. A third maximum
occurs at $-70°C$ and is attributed to hindered rotations of short main chain
segments (4 carbon atoms) in the disordered (noncrystalline) regions. Thus,
the ductility of polypropylene at $-80°C$ may be a consequence of molecular
mobility in the noncrystalline region. Two further low-temperature maxima
have been observed in polypropylene (21–23) at $-254°C$ ($19°K$) and $-220°C$
($53°K$), which have been ascribed to hindered rotations of pendent methyl
groups in the noncrystalline and crystalline regions, respectively. The $19°K$
maximum has also been ascribed to motions in disordered regions or motions
of defects, but it does not appear to be due to defect motion alone (24). Thus
the actual molecular source of these very low-temperature loss maxima is
uncertain. Whatever their molecular source the presence of mechanical loss
maxima in isotactic polypropylene at temperatures lower than $-196°C$
shows molecular mobility is available at this low temperature. This could
(and does) result in ductile failure of the polymer.

The $-80°C$ test temperature is very near to the $-70°C$ transition. The
positive skew of the true stress histograms obtained from the samples having
a spherulitic character suggests a transition in the underlying flaw process is
occurring. That is to be expected since the available modes of molecular
motion are continually being eliminated as the temperature is decreased
below a given transition region. The question is, has the transition in flaw
character progressed so far that a time-temperature superposition of data is
unreasonable at this temperature? The true stress values at $-80°C$ were all
obtained at a rate of $10^2 \%/min$. The orientation slip factor a_f has already

been determined at 23°C, so that an orientation-reduced true stress–log (a_f) slipped rate curve can be constructed for the $-80°C$ data. To bring these data onto the 23°C master curve, however, the orientation-reduced true strain must be further reduced by a temperature-reduction factor $[T(193°K)/T(296°K)]$ and then horizontally slipped along the time axis an additional amount a_T. The reduced $-80°C$ (193°K) data are shown in Fig. 4-27 before being slipped along the time axis an additional amount a_T. It is obvious that a horizontal movement along the time axis will bring this line onto the room-temperature master curve, a portion of which is also shown in the figure. The master curve in Fig. 4-25 contains the $-80°C$ reduced curve after slipping the data a log a_T of 4.432. The a_T slip factor is essentially identical to the minimum microfibrillar orientation slip factor (log $a_f = 4.477$) obtained at room temperature, which seems to agree with the true strain histogram suggestion that a microfibrillarlike stiff molecule flaw process is occurring at $-80°C$.

By $-196°C$ (77°K) the true strain histograms have a negative skew in most of the samples, suggesting that the transformation to a new flaw mechanism more characteristic of the 19 and 53°K available modes of motion is almost complete. When the true stress data are reduced by the orientation reduction factor $(1 - f_{av})$ and the temperature-reduction factor $[T(77°K)/T(296°K)]$ and are then plotted against the a_f slipped rate (Fig. 4-27) in a manner similar to that done previously with the $-80°C$ data, a horizontal line is again obtained. This indicates a microfibrillarlike flaw

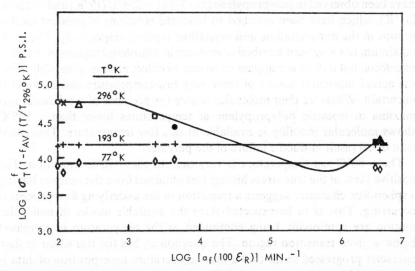

Fig. 4-27 Orientation- and temperature-reduced true stress curves for the Series A films drawn at a strain rate of 10^2 %/min at different temperatures.

process has occurred, and the true stress at break is directly proportional to the average orientation (in agreement with Fig. 4-16). The magnitudes of the reduced true stress values are too low, however, for this reduced curve to fit onto the master curve (Fig. 4-25). Thus, time–temperature superposition does not seem valid at this low temperature. Instead an extra vertical shift is required, which does not seem justified. It is interesting to note that vertical shifts are not an uncommon practice in the time–temperature super-positioning of data from crystalline polymers (ref. 16, Chapter 4, Section 11).

e. Correlation of Fracture in Fibers and Films. It was shown above how, through the use of the structural model, quantitative correlations can be obtained between the structural state of the polymer, strain rate, temperature, and the several modes of motion and flaw mechanisms available to the poly-crystalline Series A films. It is also important to apply this structural model to a spectrum of fibers and films to demonstrate its general applicability to problems of practical interest. In particular, the meaning of tenacity as a measure of fiber strength will be considered as it relates to the fracture envelope and the general problem of failure in films as well as fibers.

The tenacity of the Series D and E drawn fibers (see Chapter 3, Section C) and the Series A and B drawn films (see Chapter 3, Section B) is plotted as a function of their fabrication draw ratio in Fig. 4-28. Each point is an average

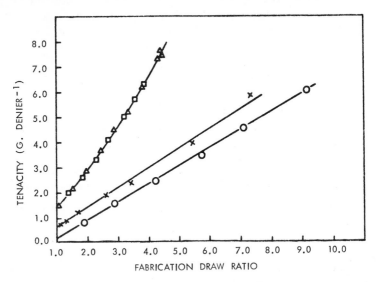

Fig. 4-28 Relation between tenacity and fabrication draw ratio for isotactic poly-propylene fibers and films drawn at a strain rate of 50%/min (5 samples each point): (O) film (draw temperature 135°C); (×) film (draw temperature 110°C); (□) fiber (draw temperature 90°C); (△) fiber (heat set).

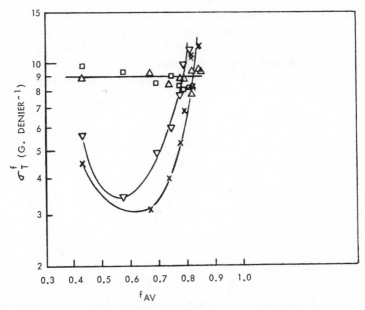

Fig. 4-29 True stress failure envelope for isotactic polypropylene fibers. Drawn fibers, strain rate: ($\square$) 50%/min, ($\triangledown$) 10^5%/min; heat-set fibers, strain rate: ($\triangle$) 50%/min, ($\times$) 10^5%/min.

of five measurements and the strain rate was 50%/min. The tensile strength of the film samples was converted to tenacity using the relation

$$\sigma^f \left(\frac{g}{\text{denier}} \right) = 0.7811 \times 10^{-4} \left[\frac{\sigma^f(\text{psi})}{\text{density (g/cm}^3)} \right]$$

Tenacity is the tensile force required to break the sample (grams) divided by a measure of mass per unit length (denier is the weight in grams of 9000 meters), and has the units of length. The strain rate of 50%/min was chosen for this study as this is typical of the slow rates of deformation generally used in fiber laboratories for standard tenacity determinations.

It was observed in the previous sections that a strain rate of 50%/min is slow enough to allow all of the film samples to transform to the microfibrillar state before fracture. That is, the true stress at break was independent of the initial orientation in the sample at this rate. The true stress at break for the Series D and E fibers is plotted against the average initial orientation in Fig. 4-29. At this typical strain rate for tenacity measurements the fibers behave as predicted from the film studies and their σ_T^f values are constant and independent of the initial orientation of the fibers.

Tenacities of the Series D and E fibers were also measured at a strain rate of 10^5 %/min to see if a fracture envelope would be obtained for fibers as it was for the films. The true stress fracture envelope for both series of fibers is shown in Fig. 4-29. The general appearance is the same as that obtained with the films. Similarly, true strain fracture envelopes for the two series of fibers are plotted in Fig. 4-30 and these also have the general appearance expected.

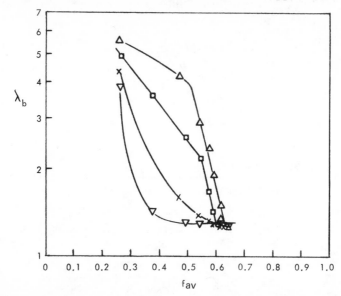

Fig. 4-30 True strain failure envelope for isotactic polypropylene fibers. Drawn fibers, strain rate: (□) 50%/min, (▽) 10^5%/min; heat-set fibers, strain rate: (△) 50%/min, (×) 10^5%/min.

The tenacity (or tensile strength) of a fiber or film is a measure of the strength of the sample at fracture with respect to the starting cross-sectional area of the sample. If the sample is strained slowly enough for the structure to be in its ultimate microfibrillar state at break, then the breaking strength will have a value controlled by the available residual extensibility due to the initial orientation state of the sample. The initial cross-sectional area of the sample is simply a function of the fabrication true strain (ln λ_f) needed to produce that initial orientation state, and ln λ_f, as can be seen from Figs. 4-5 to 4-7, is a direct function of the orientation produced during the draw. Thus, the tenacity (or tensile strength) for a slow strain rate is controlled primarily by the cross-sectional area term, which, in turn, is determined by the fabrication strain. Since the fabrication strain is proportional to the orientation, tenacity should behave similarly to the fabrication true strain in Figs. 4-5 to 4-7. The tenacity of the fibers and films, however, reduces the strength measurements of all of the samples to a common cross-sectional basis, and, thus, it would be expected that the values from both fibers and films should fall on the same general curve. Figure 4-31 shows the tenacity of all 30 fiber and film samples plotted in the manner of Fig. 4-5 (i.e., against f_{av}). All the samples are observed to fall essentially on the same curve, which has a break point of $f_{av} = 0.76$, the transition orientation from a spherulitic to a microfibrillar structure. The ultimate tenacity that could be expected from a fiber or film would occur when all the molecules are fully oriented ($f_{av} = 1.0$) and the value obtained by extrapolation of the curve is 15 g/denier. From Figs. 4-5 to 4-7, it would be predicted that a plot of the tenacity against the average initial amorphous orientation in the sample (f_{am}) should be a straight line with no break in the curve. The tenacity is plotted according to Fig. 4-7 in Fig. 4-32. A straight line is obtained as predicted, which again extrapolates to an ultimate tenacity of 15 g/denier. Attempts to produce special supertenacity fibers (25) resulted in several samples with a tenacity of 12 g/denier and one sample with a tenacity of 13 g/denier. Considering the impossible task of obtaining a perfectly oriented fiber, these experimental results are in excellent agreement with the extrapolated prediction of tenacity (max) = 15 g/denier. Thus, by utilization of quantitative structural criteria, it is now possible to understand and bring together a large body of fracture results from fibers and films into a meaningful coherent pattern.

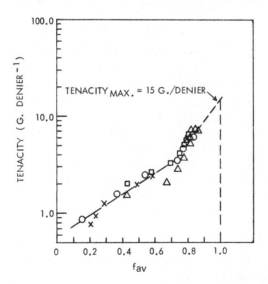

Fig. 4-31 Comparative tenacities of fibers and films as a function of the average fabricated molecular orientation. Tensile strain rate is 50%/min (5 samples for each point): (○) film (draw temperature 135°C); (×) film (draw temperature 110°C); (□) fiber (draw temperature 90°C); (△) fiber (heat set).

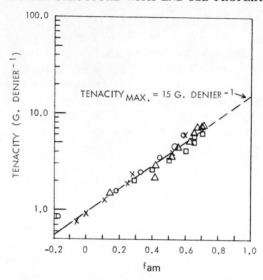

Fig. 4-32 Comparative tenacities of fibers and films as a function of the amorphous orientation. Tensile strain rate is 50%/min (5 samples for each point): (O) film (draw temperature 135°C); (×) film (draw temperature 110°C); (□) fiber (draw temperature 90°C); (△) fiber (heat set).

3. Yield Behavior and Initial Structure

When a polymer undergoes yielding it is called ductile. When it fractures before yielding, it is called brittle. In this study all of the Series A films (80 samples per measurement) having a spherulitic structure showed both upper and lower yield points at all of the strain rates and temperatures examined. Thus, these samples were ductile under all experimental conditions. The films having a microfibrillar structure, on the other hand, were essentially brittle. At a strain rate of 10^2 %/min the $f_{av} = 0.746$ microfibrillar samples yielded; however, only 32% of the highest oriented samples ($f_{av} = 0.792$) showed yield behavior at this strain rate. Neither of these samples showed any yields at a strain rate of 10^3 %/min, but both showed a few yields at 10^5 %/min (ten of the $f_{av} = 0.746$ samples yielded, and three of the $f_{av} = 0.792$ samples). Under all other conditions brittle failure was observed for the microfibrillar samples. Thus, for the samples studied, there is a ductile–brittle transition in going from a spherulitic structure (ductile) to a microfibrillar structure (brittle).

The ductile behavior of the spherulitic samples occurs as a consequence of their internal structure. The spherulitic structural region is one in which large crystallites and poorly oriented noncrystalline material is present, and shear forces can align the molecular chain axis in these structures into the deformation direction. It should be realized that not only is the noncrystalline polymer ductile, but the polymer crystals are quite ductile as well. Thus, lamellar slip has been observed in polypropylene single crystals (26), and mats of ethylene–propylene copolymer single crystals have been cold drawn to 500% extension before fracture (27). The microfibrillar structure contains small, highly oriented crystallites, strained, highly oriented noncrystalline chains, and voids between micrifibrils. This type of structure would not be expected to manifest ductile failure.

The effect of orientation on the general concept of a brittle–ductile transition temperature is also significant. The fracture and yield stresses for polypropylene have been reported as a function of temperature (28). Above about −10°C the failure was reported as ductile (i.e., there is yielding before fracture), whereas below −10°C the failure was brittle (no yield). The point at which the brittle failure stress versus temperature line intersected the yield stress versus temperature line was called the brittle–ductile transition temperature (−10°C). The structural state of the sample that was studied was not reported. The yield stress for the spherulitic Series A samples of isotactic polypropylene, deformed at a strain rate of 10^2 %/min at different temperatures, is shown in Fig. 4-33. Since all of the samples yielded at all

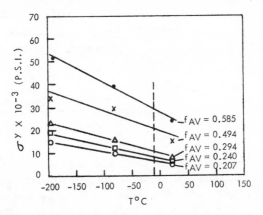

Fig. 4-33 Effect of temperature and average orientation on the yield stress of isotactic polypropylene film (strain rate: 10^2 %/min).

temperatures, there is no brittle–ductile transition down to −196°C, and the yield stress is a linear function of the temperature. The only samples which showed a brittle–ductile transition were the microfibrillar samples and then only at a rate of 10^2 %/min at room temperature. (At all other rates and temperatures, failure was brittle only.) Thus the concept of a brittle–ductile transition temperature is not fruitful in this study, since the transition in behavior seems to be structure and not temperature dependent.

As only minor structural rearrangements occur in the microstructure before the activation energy for the yield process is reached, the yield stress σ^y (Fig. 4-11), would be expected to be some function of the original structure in the sample. The strain rate and initial average orientation dependence of the yield stress σ^y for the Series A films strained at room temperature, is shown in Fig. 4-34. The yield data observed at a strain rate of 10^6 %/min at room temperature showed the same trend in yields as that observed at 10^5 %/min, but noise problems prevented quantitative evaluation of the data. The yield stress is seen to be a linear function of log (strain rate) for each orientation (the sample numbers are listed on the right ordinate of the figure) and increases with increasing orientation and increasing rate of deformation. The yield stress dependence on the initial orientation in the sample can be seen more clearly from Fig. 4-35. Here log σ^y is a linear function of the

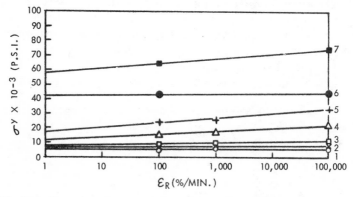

Fig. 4-34 The effect of strain rate ϵ_R and orientation on yield strength, σ^y at 23°C. f_{av} for the representative numbers are (1) 0.207, (2) 0.240, (3) 0.294, (4) 0.494, (5) 0.585, (6) 0.746, (7) 0.792.

average initial orientation in the sample for all rates of deformation and temperatures studied. The yield stress is again seen to increase with increasing orientation, increasing strain rate, and with decreasing temperature as well. Thus the initial orientation in the sample is of primary importance in determining the yield stress.

The dependence of the yield stress on strain rate, initial structure, and temperature suggest that a yield stress master curve could be constructed using both structural- and temperature-reduced variables. Figure 4-36 shows the logarithm of the yield stress at 23°C multiplied by the structure reduction factor $(1 - f_{av})$, plotted against the logarithm of the percent strain rate for the Series A films. The orientation reduction factor has lowered the log σ^y curves of Fig. 4-34 so that they overlap. These curves are then slid along the time axis until they form the master curve shown in Fig. 4-37. The

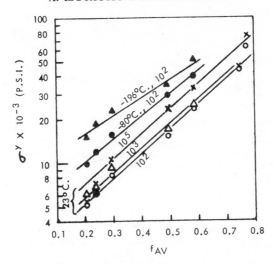

Fig. 4-35 Effect of average orientation, strain rate, and temperature on the yield stress of isotactic polypropylene film.

amount of reciprocal time each curve was slipped along the time axis is recorded on the figure as the orientation slip factor for yielding a_f^y. The figure includes a few yield values available from the microfibrillar samples. The fit of the points is good and indicates an increase in yield strength with increasing rate (i.e., at higher rate the molecules are stiffer and harder to deform) and increasing initial orientation, as already shown in Figs. 4-34 and 4-35.

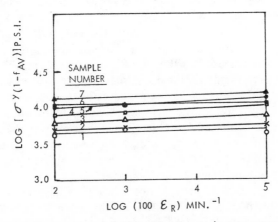

Fig. 4-36 Structure-factor reduced yield stress versus strain-rate curves for the Series A films at room temperature. f_{av} for the representative numbers are: (1) 0.207, (2) 0.240, (3) 0.294, (4) 0.494, (5) 0.585, (6) 0.746, (7) 0.792.

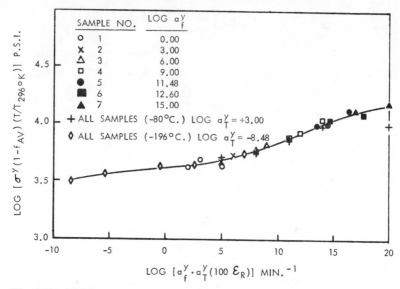

Fig. 4-37 Yield stress master curve for the Series A films.

Time–temperature superposition of the yield stress data was also tried in the manner utilized to obtain the true stress failure master curve (Fig. 4-25). As with the true stress to failure results, the yield stress data obtained at −80°C fitted the master curve after reduction for structure and temperature. The temperature slip factor for yielding was $\log a_T^y = +3.00$. Thus, lowering the temperature moves the yield stress in the same direction as increasing the orientation or increasing the rate. This is the expected behavior from a reduction in temperature.

The reduced yield stress data for deformation at −196°C is included on the curve for the purpose of illustration. From Fig. 4-35 the yield stress at −196°C is seen to increase over that at −80°C for all samples. The magnitude of the increase is too small for the decrease in temperature, however. That is, multiplication of the orientation-reduced data by the temperature reduction factor ($T_{77°K}/T_{296°K}$) lowers the yield stress values below those of the low-orientation, room-temperature values. Thus, the only way these data could be fitted to the master curve was to slip the curve toward longer times and lower orientation (i.e., $\log a_f^y = −8.48$)—a direction inconsistent with the general theory of time–temperature behavior.

Thus, the −196°C yield data do not behave in a manner consistent with the other temperature and orientation results. Such an inconsistency was also observed with the fracture behavior, which suggested a new mode of failure was occurring. As −196°C is near the low-temperature (19 and 53°K) transitions, a new mode of yield behavior might also be expected.

Studies of yielding in poly(methyl methacrylate) (29) have shown, by simple shear and uniaxial tension yielding experiments, that the shear stress required for yielding is lowered by the presence of a dilatational stress. Examination of motion pictures taken of a few of the Series A films during deformation at $-196°C$ indicates much of the observed deformation in these samples was dilatational. Thus the observed lower-yield stress at $-196°C$ is consistent with the observed dilatation and suggests that new modes of yielding are contributing to the behavior at this low temperature.

The orientation slip factor for yielding, log a_f^y, would be expected to be a function of the initial orientation in the sample. Log a_f^y is plotted against the average initial sample orientation in Fig. 4-38. A reasonable linear

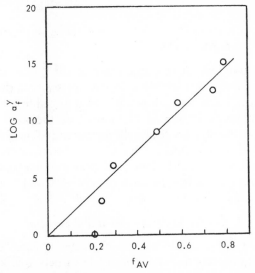

Fig. 4-38 Relation between the orientation slip factor for yield a_f^y and the average initial orientation f_{av} in the Series A films.

relation is obtained. It is interesting to note that the two a_f^y values from the microfibrillar samples fit the best line between the data points. Thus the few observed microfibrillar yields give values consistent with the general treatment. Again, a quantitative definition of the structural state of the polymer has made it possible to correlate a large amount of mechanical data.

The initial state of orientation of the polymer will have a strong influence on the yield strain ϵ_y as well as the yield stress, as can be seen from Fig. 4-39. The yield strain initially increases with increasing orientation in the spherulitic structural region. As the transition region between a spherulitic and a microfibrillar structure is approached the yield strain begins a gradual decrease, then drops precipitously in the microfibrillar region. Since yielding in the

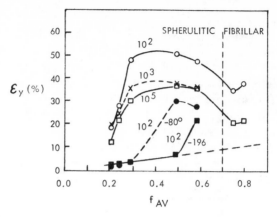

Fig. 4-39 Effect of temperature, strain rate, and average orientation on the yield strain of isotactic polypropylene films.

microfibrillar region (except at a strain rate of $10^2\%/min$) is not generally representative of microfibrillar behavior, this precipitous drop in ϵ_y is expected and, in fact, could more reasonably be dropped to zero.

Further examination of Fig. 4-39 shows that the yield strain is also influenced both by the strain rate and the temperature. The greater the strain rate the lower will be the yield strain. Also the lower the temperature the lower will be the yield strain. The high yield strains attainable by the oriented spherulitic samples at the low temperatures of -80 and $-196°C$ are particularly noteworthy.

4. Tensile Recovery and Initial Structure

Tensile recovery is generally used to characterize fibers because of the similarity of this property to fiber behavior in carpets and fabrics. When a carpet is stepped on, the fibers will be deformed by a small strain. After the foot is removed, the fibers in the carpet will begin to recover their original position. In a good carpet the recovery will be almost complete, and will occur in a reasonable time span. Two types of tensile recovery measurements can be obtained from a single recovery curve. A schematic representation of the recovery curve is shown in Fig. 4-40. Initially the sample is stretched slowly

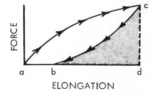

Fig. 4-40 Schematic representation of a tensile recovery curve.

(here 50%/min) to a given extension (in this study 5%). This is represented by curve ac. When the desired extension has been reached, the crosshead direction is reversed and the sample is returned to its starting position. If the material is not ideally elastic the return stress–strain curve (curve cb in Fig. 4-40) will produce a hysteresis loop. The elastic recovery for the fiber R_E is defined as the fractional length of the fiber recovered from the initial extension. That is, $R_E(\%) = (bd/ad) \times 100$. This recovery parameter is important because in a carpet or fabric it is really strain recovery that interests us. However, the storage of work is needed to overcome interfiber friction in the recovery process. Thus, in order to recover their original position in a carpet or fabric, the fibers must slide past each other, and work is expended in overcoming the frictional drag. A work-recovery term can also be obtained from the tensile-recovery measurement. The work recovery R_W is defined as the area of the stress–strain curve (work energy) recovered from that required for the initial extension, that is, in Fig. 4-40, $R_W(\%) =$ (area bcd/area acd) $\times$ 100.

Both the elastic and work recovery observed in a fiber or a film will depend on the manner in which the sample was fabricated. The work recovery and elastic recovery, respectively, of the Series A and B films and Series D and E fibers are shown in Figs. 4-41 and 4-42 plotted against the fabrication draw ratio. The curves have little in common except that the magnitude of the changes in work recovery with fabrication draw are more severe than those of elastic recovery. Since it is the combination of the two properties and not the

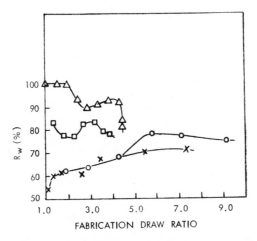

Fig. 4-41 Work recovery as a function of fabrication draw ratio for a 5% room-temperature extension cycle of isotactic polypropylene fibers and films: (O) film (draw temperature 135°C); ($\times$) film (draw temperature 110°C); ($\square$) fiber (draw temperature 70°C); ($\triangle$) fiber (heat set).

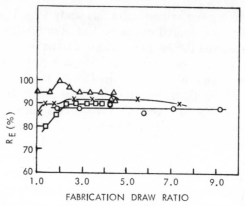

Fig. 4-42 Elastic recovery as a function of fabrication draw ratio for a 5% room-temperature extension cycle of isotactic polypropylene fibers and films: (O) film (draw temperature 135°C); (×) film (draw temperature 110°C); (□) fiber (draw temperature 90°C); (△) fiber (heat set).

individual properties themselves that are important in carpet and fabric behavior, a new parameter can be defined that is the product of the work and elastic recovery. That is, the recovery factor $(\%) = (R_E R_W)100$. The behavior of the recovery factor with fabrication draw ratio is shown in Fig. 4-43. Again no obvious relationships can be observed for samples produced by different fabrication conditions. All of the curves manifest at least one

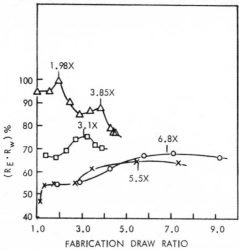

Fig. 4-43 Recovery factor as a function of fabrication draw ratio for a 5% room-temperature extension cycle of isotactic polypropylene fibers and films: (O) film (draw temperature 135°C); (×) film (draw temperature 110°C); (□) fiber (draw temperature 90°C); (△) fiber (heat set).

maximum in recovery factor with draw ratio. The draw ratios of the respective maxima are listed in the figure, and again there is no obvious connection between the position of the maxima and the fabrication conditions.

The different fabrication conditions of the film and fiber processes are simply different paths by which a given average orientation is produced in the sample (Fig. 4-5). The proper parameter with which to examine the recovery behavior should, therefore, be a quantitative measure of the

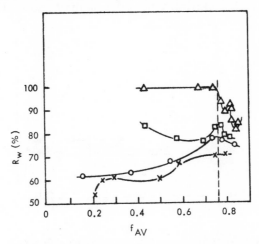

Fig. 4-44 Work recovery as a function of average orientation for a 5% room-temperature extension cycle of isotactic polypropylene fibers and films: (O) film (draw temperature 135°C); (×) film (draw temperature 110°C); (□) fiber (draw temperature 90°C); (△) fiber (heat set).

average structure in the sample and not the fabrication draw ratio. The work recovery, the elastic recovery, and the recovery factor of the fibers and films are plotted against the average initial orientation in the samples in Figs. 4-44 to 4-46, respectively. The primary maximum of all of the samples now occurs at the same f_{av} value of 0.76. This is the region of the spherulitic to microfibrillar structural transition. Thus the recovery of the sample improves with an increase in the orientation of the interconnected intraspherulitic substructure, as permanent deformation of this coherent tough mass becomes increasingly difficult. Once the microfibrillar structure has been produced, its recovery problems become similar to those common to a fabric or carpet. That is, the microfibrillar substructure not only has its own structural strength problems but the microfibrils can now also deform by movement of fibrils past each other. Thus, there are more potentially permanent deformation mechanisms available to the microfibrillar structure than are available to the spherulitic structure.

There is a second maximum in the heat-set fiber samples, which occurs at a higher average orientation than the primary maximum (Fig. 4-46). The position of this maximum seems to correspond to the second maximum observed in the long-spacing curve for this fiber (Fig. 4-9). This suggests the strained noncrystalline chains are contributing to the result. Figure 4-47 shows the recovery factor plotted against the noncrystalline chain orientation function f_{am}. The recovery-factor curves are now found to fall into two sets of maxima having different relative positions on the f_{am} scale. The significance

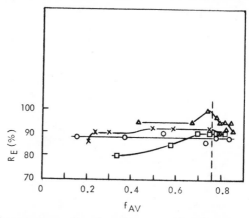

Fig. 4-45   Elastic recovery as a function of average orientation for a 5 % room-temperature extension cycle of isotactic polypropylene fibers and films: (○) film (draw temperature 135°C); (×) film (draw temperature 110°C); (□) fiber (draw temperature 90°C); (△) fiber (heat set).

of this plot is simple. The fibers drawn at 90°C had not completely fibrillated (8) due to the low draw temperature (Fig. 4-6). Instead the noncrystalline chains became highly strained during the deformation. The heat-set fibers required a high draw to get an equivalent high strain in the noncrystalline chains. Thus, the primary maximum in the drawn fibers and the secondary maximum in the heat-set fibers are both a consequence of the same high noncrystalline strain. The other group of three primary maxima are controlled by the crystalline region rather than by the noncrystalline region.

All of the samples, however, have their primary maximum occurring at the transition between a spherulitic and a microfibrillar structure, with the recovery factor decreasing with microfibrillation. Since the recovery factor is essentially a strain measurement, it is interesting to note the similarities in behavior between the recovery factor and the yield strain curves (Fig. 4-39). That is, both curves go through a maximum and the maximum occurs near the spherulitic-to-microfibrillar transition orientation. The deformation to yielding is much more severe than the 5 % deformation of the tensile recovery

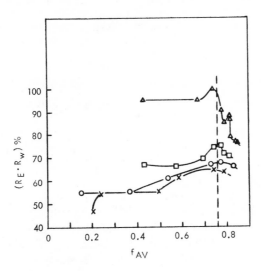

Fig. 4-46 Recovery factor as a function of average orientation for a 5% room-temperature extension cycle of isotactic polypropylene fibers and films: (O) film (draw temperature 135°C); (×) film (draw temperature 110°C); (□) fiber (draw temperature 90°C); (△) fiber (heat set).

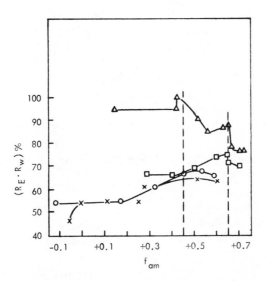

Fig. 4-47 Recovery factor as a function of amorphous orientation for a 5% room-temperature extension cycle of isotactic polypropylene fibers and films: (O) film (draw temperature 135°C); (×) film (draw temperature 110°C); (□) fiber (draw temperature 90°C); (△) fiber (heat set).

test, but the recovery test shows that even at these low extensions some permanent deformation has occurred in the structure.

5. Summary

In order to develop a fundamental understanding of the effects of time and temperature on the physical behavior of a material, it is necessary to have experimental criteria to define its structural state. By examining the changes that occur in the material's state with changes in time and temperature, it is then possible to develop structural models that describe the mechanics of the processes that occur. The availability of quantitative structural information has made possible an examination of fabrication, fracture, yielding, and recovery behavior of both fibers and films of isotactic polypropylene as a single general problem. This has resulted in new quantitative relationships between mechanical properties and morphological structure and has developed a deeper insight into the underlying mechanisms that are occurring. The general conclusions of this section can be summarized as follows:

1. Mechanical properties of isotactic polypropylene fibers and films can be quantitatively correlated with polymer structure.

2. The average state of orientation in the polymer is a useful expression for comparing structure–property relationships.

3. Fabrication processes can be considered as different paths for reaching equivalent average orientation states of a material.

4. Fabrication true strain is linearly related to the final fabricated orientation.

5. A true stress fracture envelope can be constructed, which quantitatively correlates failure as a function of rate and temperature in mechanistically meaningful morphological terms.

6. A true strain fracture envelope can be described in a manner similar to 5 (above).

7. Failure histograms suggest different flaw distributions cause failure at slow and fast strain rates at room temperature and at a very low temperature $(-196°C)$.

8. A true stress master fracture curve can be constructed, which includes time, temperature, and the structural state of the material.

9. The true stress master curve slip factor is a direct function of the structural state of the polymer.

10. Yield stress increases with increasing molecular orientation, increasing rate, and decreasing temperature.

11. Yield stress correlates directly with average molecular orientation.

12. A yield stress master curve can be constructed, which includes time, temperature, and the structural state of the material.

13. The yield stress master curve slip factor is a direct function of the structural state of the polymer.

14. Isotactic polypropylene is ductile (has a yield point) from $-196°C$ to room temperature for all of the samples examined that have spherulitic character, but manifests brittle behavior for the samples having a microfibrillar structure.

15. Yield strain increases with increasing orientation for spherulitic samples but decreases with microfibrillation.

16. Yield strain decreases with increasing strain rate and decreasing temperature.

17. Tensile recovery reaches a maximum in fibers and films at the transition region between a spherulitic and a microfibrillar structure.

B. POLY(ETHYLENE TEREPHTHALATE)

1. Introduction

Poly(ethylene terephthalate) (PET) is another polycrystalline polymer. It, like isotactic polypropylene, also has structural order on the molecular and interlamellar levels. The character of the structural order varies with the conditions of fabrication. Certainly, the mechanical properties of a sample of PET depend on its particular structural arrangement.

Realizing that the observed mechanical properties of a polycrystalline polymer are intimately related to the internal morphological structure of the polymer, the character of these relationships was examined for isotactic polypropylene in the last section. There it was demonstrated that large amounts of mechanical data could be simply and quantitatively correlated through the use of structural criteria. Are the intimate relationships between structure and properties observed for polypropylene specific for that polymer, or are they, as would seem more reasonable, characteristic of polycrystalline polymers in general? The purpose of this section is to demonstrate that the morphological model developed for polypropylene is a general one. This is accomplished by examining the structure–property relations of PET (a crystalline polymer very different from isotactic polypropylene, for isotactic polypropylene was 60–70% crystalline, while PET is 60–70% noncrystalline and glassy). Application of the PET data in terms of a general structural model will be shown to lead directly to simple, quantitative relationships between the structural state of PET and such properties as tenacity, shrinkage, dynamic loss modulus, tensile modulus, long spacing, and the intensity from small-angle x-ray scattering.

The simple, direct, and logical manner in which structure relates to mechanical properties was demonstrated in the previous section on isotactic polypropylene fibers and films (30). In that study the properties of 30 different structural states produced by four different fabrication processes were examined. The different fabrication processes were shown to produce equivalent structural states, and the mechanical properties of the samples were then shown to depend not on the fabrication process, but on the structural state of the polymer at the time of the test. In fact, the fabrication process and the test procedure could be considered as simply different deformation processes leading to some given structural state of the polymer.

The orientation-state model of a material should be independent of the polymer under examination. The same structural definitions will apply to PET as apply to isotactic polypropylene and thus structure–property relations of a generally similar form should be expected for both polymers. The structural relations examined for polypropylene fell into two main categories: (1) those directly related to the average orientation of the sample, f_{av} [here $f_{av} = \beta f_c + (1 - \beta) f_{am}$], and (2) those directly related to the noncrystalline orientation f_{am}. The modulus, the failure envelope, the yield behavior, and the resilience were all found to be directly related to the average orientation state of the polymer. On the other hand, the true fabrication strain (which is defined as $\ln \lambda_F$, where λ_F is the fabrication draw ratio) was found to be a linear function of the noncrystalline orientation function f_{am}, and the tenacity of those samples that were drawn at a slow enough rate to allow maximum orientation to occur before failure was also directly proportional to the noncrystalline orientation function.

Tables 4-1 and 4-2 contain both the structural data required to define the structural state of the PET samples and the following mechanical data: tenacity, shrinkage, fabrication draw ratio, tensile modulus, and the dynamic loss modulus E''. Also included are the crystal long-spacing and small-angle x-ray intensity of the fibers (31–33). With the structure–property relations determined for isotactic polypropylene as a guide, the data from each of these measurements are examined structurally below. In this way the general character of the observed structure–property relations are demonstrated, and any deviation of the PET data from the general form of the structural relations developed for polypropylene becomes obvious.

Before examining the structure–property correlations for PET it is helpful to understand how the PET samples were prepared and the conditions under which they were tested. The procedures used for the preparation and testing of the PET fibers were as follows (31–33):

Spun PET of low crystallinity (less than 2% by x-ray measurements) and of birefringence 0.002 was drawn to draw ratios up to 5× over a hot pin at 80°C. Yarns of 3× and 5× draw were used in the annealing experiments. Two types of annealing procedures were used: (1) 5-m lengths of both the 3× and 5× drawn yarn, in a wire mesh basket, were immersed in hot silicone

Table 4-1. Mechanical and Structural Data for PET Yarns[a]

Sample	Crystallinity, %	Birefringence	Crystallite orientation function, f_c	f_{am}	Shrinkage, %	Modulus, g/denier	Tenacity, g/denier	Long period, Å	Intensity in Small-Angle Maximum
3 × PET									
in oil at 20°C	12	0.140	0.915	0.505	—	—	2.9	—	—
100°C	16	0.127	0.915	0.433	11	37	3.6	132	0.5
150°C	30	0.149	0.915	0.485	19	27	3.1	120	56
Crimped, at 175°C	37	0.150	0.889	0.472	24	26	3.0	127	77
200°C	38	0.142	0.915	0.400	33	17	2.7	144	115
225°C	36	0.132	0.858	0.382	52	11	1.7	169	155
240°C	24	0.145	—	—	67	—	—	192	50
5 × PET									
in oil at 20°C	35	0.206	0.943	0.790	—	—	5.4	—	—
100°C	38	0.200	0.943	0.811	8	66	6.7	124	0.8
150°C	34	0.174	0.910	0.709	23	48	5.8	122	47
175°C	38	0.164	0.910	0.668	31	35	4.8	124	76
200°C	39	0.159	0.858	0.680	43	27	3.7	147	147
Crimped, at 225°C	39	0.112	0.858	0.480	60	12	2.2	167	214
240°C	33	0.040	—	—	75	14	1.9	195	132
5 × PET									
in air at 100°C	31		0.93	0.84	8	77	7.1	132	13
150°C	34		0.93	0.75	20	51	5.8	120	52
175°C	40		0.93	0.75	24	44	5.3	120	76
200°C	38		0.92	0.73	30	46	4.6	132	127
225°C	38		0.92	0.68	35	32	3.7	140	172
240°C	40		0.93	0.68	44	36	4.1	151	212

[a] Data of Dumbleton (32,33).

Table 4-2. Vibron and X-Ray Data for 3 × and 4.25 × Drawn PET Yarns Annealed for 6 hr Followed by Treatment in Boiling Water[a]

Sample	Treatment, Temp, °C	Crystallinity, %	f_c	f_{am}	Temperature of E''_{max}, °C
3 ×	No heat	33	0.91	0.53	126
	100	33	0.915	0.54	128
	125	31.5	0.920	0.54	130
	150	38	0.91	0.57	128
	175	43	0.92	0.57	123
	200	56	0.92	0.56	110
	220	57	0.91	0.56	110
	240	59	0.91	0.54	100
4.25 ×	No heat	38	0.92	0.68	147
	100	40	0.93	0.68	150
	125	42	0.92	0.64	153
	150	45	0.92	0.67	142
	175	47	0.93	0.69	140
	200	51	0.94	0.72	140
	220	53	0.93	0.73	124
	240	57	0.93	0.70	120

[a] Data of Dumbleton et al.[31]

oil for 1 min; (2) 2.5-m lengths of the 5 × drawn yarn were immersed in a test tube of heated air for 10 min. These annealing times were chosen to obtain maximum shrinkage. All samples were quenched in carbon tetrachloride at room temperature after annealing was completed. The shrinkage ratio S was calculated from the formula

$$S = \frac{\text{initial length} - \text{final length}}{\text{initial length}} \tag{4-6}$$

Crystallinity measurements were made with a Norelco x-ray diffractometer, which was also used to obtain the crystallite orientation from scans on the $(\bar{1}05)$ reflection. The orientation parameter employed was f_c given by eq. 2-9 (5):

$$f_c = (\tfrac{1}{2})(3\langle\cos^2\theta\rangle - 1) \tag{2-9}$$

where $\langle\cos^2\theta\rangle$ is the mean-square cosine of the angle θ between a $(\bar{1}05)$ plane normal and the fiber axis.

Small-angle x-ray data were obtained with a Kratky camera (slit collimation). Integrated intensities of the long-period maximum were derived from the areas under the maximum after a linear background had been subtracted.

The tensile modulus and tenacity were determined on an Instron testing machine at 70°F and 65% RH with single filaments.

Birefringence measurements were made on single filaments by using a compensator. Because of nonuniformity in denier, it was difficult to get consistent measurements for the 3× yarns.

The intrinsic birefringences for PET ($\Delta_c^0 = 0.220$, $\Delta_{am}^0 = 0.275$) were determined by Dumbleton using the method of Samuels (1,5). By combining the intrinsic birefringences Δ_c^0 and Δ_{am}^0, the crystal orientation function f_c, the fraction of crystals β (from the percent crystallinity), and the measured birefringence Δ_T, through the equation (5):

$$\Delta_T = \beta \Delta_c^0 f_c + (1 - \beta) \Delta_{am}^0 f_{am} \tag{2-36}$$

it was then possible to calculate f_{am}, the orientation function for the non-crystalline chains.

The modulus and tenacity values listed in Table 4-1 for the 3× drawn yarns were not listed in the references but were taken from plotted values in ref. 32.

The loss modulus E'' data listed in Table 4-2 were obtained from samples treated as follows (31):

PET yarns drawn 3× and 4.25× over a hot pin at 80°C were employed. Samples were heated for 6 hr at temperatures up to 240°C *in vacuo* under conditions in which the samples were free to shrink. All samples were subsequently boiled in water, under relaxed conditions, for 1 hr.

Dynamic measurements were made with a Vibron direct reading viscoelastometer operated at a frequency of 11 Hz. Samples were heated at 1°C/min in a nitrogen atmosphere under relaxed conditions; and measurements of the tensile modulus E' and the loss factor tan δ were made at increments of 5°C, except near the transition region when smaller increments were used. The loss modulus E'' was calculated from the relation $E'' = E' \tan \delta$.

The crystallinity, crystallite orientation, and birefringence of these samples were obtained by using the procedures described in Chapter 2.

2. Structure–Property Correlations

a. Tenacity.   When a cast film or unoriented fiber is deformed, the randomly oriented molecules tend to become oriented in the deformation direction. If the sample is deformed slowly enough and at a sufficiently high temperature it will follow the stress–strain path shown in Fig. 4-48. Under these conditions the molecules have enough mobility and time available to relieve the applied stress by rearrangement of the internal structure of the sample. Opposed to this plastic deformation of the sample is the presence

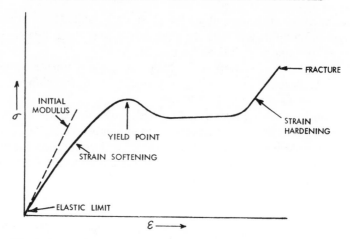

Fig. 4-48　Schematic representation of a stress–strain curve.

within the sample of a distribution of flaws (stress concentration regions), which, under the proper deformation conditions, can cause crack formation and ultimate sample failure. As the deformation process proceeds through the yield point and into the strain-hardening region of the stress–strain curve, stress builds up in the highly oriented structure until no further plastic deformation mechanisms are available to the molecules. At this point the flaw mechanism predominates and sample failure occurs.

Tenacity is a measure of the failure strength of a fiber. It is defined as the tensile stress required to break the sample (grams) divided by a measure of mass per unit length (denier is the weight in grams of 9000 meters), and thus tenacity has the units of length. Tenacity represents the tensile stress measured at break divided by the original dimensions of the fiber before deformation.

Studies of the failure envelope of isotactic polypropylene demonstrated that, when the rate of deformation at a given draw temperature is slow enough for the molecules to reach the highest orientation possible before the flaw mechanism predominates, the force required for failure of a given cross-sectional area of the sample will be a constant (Section A.2.b). That is, the true stress at break (the strength per breaking cross-sectional area of the sample) is independent of the starting orientation of the sample. This is true because the molecules in all of the samples are equally oriented at the time of break irrespective of the initial orientation state. Under these conditions of constant true breaking stress, the tenacity (which measures the breaking force in terms of the unstretched dimensions of the sample and not in terms of the dimensions at the time of break) is simply a measure of the dimensional change available to the sample before failure occurs. Thus, under these special conditions, tenacity is essentially a reciprocal measure of the extensibility (since the smaller the extensibility range the greater the tenacity) normalized to the dimensions of the starting sample (Section A.2.e).

The tenacity of the drawn and annealed PET samples are plotted as a function of annealing temperature in Fig. 4-49. The tenacities of both the

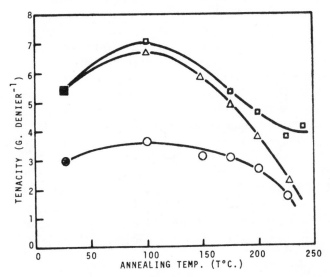

Fig. 4-49 Relation between the annealing temperature and the tenacity of PET fibers: (△) 5× oil-annealed fibers; (□) 5× air-annealed fibers; (○) 3× oil-annealed fibers; (■, ●) unannealed drawn fibers.

3× and the 5× yarns decrease with annealing temperature after an initial increase over that of the original draw yarn. Except for the general similarity in shape between the curves for the differently treated yarns, and the fact that the tenacities of the 3× yarns are lower than those of the 5× yarns, there seems to be little that can be correlated.

Considering these same tenacity data from a structural point of view, we see a different picture emerging. Free shrinkage occurs primarily as a consequence of the relaxation of the strained, oriented, noncrystalline chains. This will become obvious when shrinkage is examined in the next section. The greater the annealing temperature the greater the molecular relaxation possible. Since the tenacity value depends on the distance of the starting orientation state from the maximum value allowable at the temperature of the tenacity test, lower tenacities will be expected from samples relaxed to lower orientation states.

The tenacity studies on isotactic polypropylene showed that the controlling structural feature for tenacity relations was the noncrystalline orientation. When log tenacity was plotted against f_{am}, the tenacities obtained from 30 structural states, fabricated by four different processes, all fell on a single line (Section A.2.c). The primary experimental requirement here was that the tensile draw rate be slow enough for all of the samples to reach the same limiting structure before failure. An extension rate of 50%/min or less was required for the polypropylene samples to satisfy this criterion.

A strain rate of 50%/min is typical of the slow rates of deformation generally used in fiber laboratories for standard tenacity determinations. The rate of deformation used to determine the tenacity of the PET yarns was not reported (32,33). The extension ratio at break was also missing (32,33). Without the extension ratio at break the true stress cannot be calculated and, thus, the independence of the true stress at failure with sample structure cannot be checked. However, since the standard rate of deformation of tenacity samples is slow and the limiting rate of strain for PET is not known, it is reasonable to assume that the experimental PET measurements were made in the desired constant true stress region. If this is so then the study of isotactic polypropylene would predict that a semilog plot of tenacity against f_{am} would be linear, with all the PET samples fitting the same line.

Figure 4-50 shows a semilogarithmic plot of tenacity against f_{am}. All the points fall around a single line, as predicted. These points include the original $3 \times$ and $5 \times$ drawn samples (solid filled points), each of the original drawn samples after free shrinkage in oil at different temperatures, and

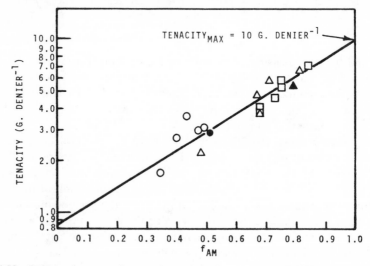

Fig. 4-50 Relation between the tenacity and f_{am} for annealed PET fibers: ($\triangle$) $5 \times$ oil-annealed fibers; ($\square$) $5 \times$ air-annealed fibers; ($\bigcirc$) $3 \times$ oil-annealed fibers; ($\blacksquare$, $\bullet$) unannealed drawn fibers.

the 5 × drawn sample after free shrinkage in air at different temperatures. All of these samples were then further drawn during the tenacity measurement until they failed. The correlations obtained in Fig. 4-50 between the orientation state of the noncrystalline chains before the tenacity test and the final measured tenacity, show that the rate of deformation during the tenacity test was slow enough for all of the samples to fail under the same structural conditions irrespective of their starting structure. It also shows that under these conditions the measured tenacity of the fibers depends on the noncrystalline orientation state of the fiber at the time of the test and not on the particular fabrication history required to reach that structural state. In this sense the structural dependence of tenacity for PET is the same as that previously observed for isotactic polypropylene.

Once a quantitative correlation between tenacity and structure has been established it is possible to predict the maximum tenacity to be expected from a fully oriented PET fiber. This is done by extrapolating the tenacity versus f_{am} line to a value of $f_{am} = 1.0$. Extrapolation of the line in Fig. 4-50 leads to a maximum tenacity of 10 g/denier. An experimental tenacity as high as 9.6 g/denier has been reported (34) for PET.

b. Thermal Annealing.

(1) *Shrinkage.* When the essentially unoriented noncrystalline PET spun fiber is drawn to different extensions at 80°C, two processes occur: (1) the noncrystalline chains become oriented and (2) crystallization occurs. If these drawn fibers are then subjected to increased temperature and are allowed to shrink freely for a fixed time, two further processes occur; (1) the orientation of the noncrystalline chains decreases and (2) the crystallites thicken.

What is the relation between the structures developed during the drawing processes and the subsequent shrinkage process? Is the observed shrinkage simply related to orientation-relaxation processes? Is there a fundamental structural difference between the shrinkage mechanism that occurs in oil at different temperatures and the shrinkage mechanism that occurs in air? Can characterization of the shrinkage phenomena be placed on a quantitative structural foundation?

The effect of annealing temperature and medium on the shrinkage of the drawn PET fibers is shown in Fig. 4-51. The shrinkage is seen here to increase with increasing temperature in both oil and air, with the fibers annealed in oil showing a significantly greater shrinkage at higher temperatures than those annealed in air.

The fibers in question are highly amorphous (ca. 60–70%) and the amorphous chains are oriented. Though the glass transition temperature of unoriented amorphous PET is 67°C, that of the oriented PET can vary

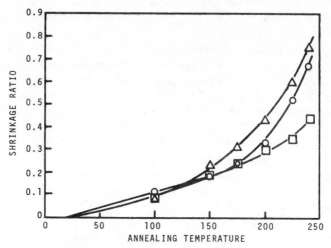

Fig. 4-51 Relation between the annealing temperature and the shrinkage ratio of PET fibers: (△) 5 × oil-annealed fibers; (□) 5 × air-annealed fibers; (○) 3 × oil-annealed fibers.

from 100°C up to at least 160°C, depending on the amount of orientation (see Table 4-2 and Section B.2.c). Thus the annealing temperatures used to shrink the fibers are in the correct range for imparting mobility to the oriented noncrystalline chains. The amount of disorientation of the non-crystalline chains will depend on the annealing temperature and the time. The greater the annealing temperature for a given time in a given medium, the greater would be the expected disorientation and, hence, the greater would be the shrinkage.

The shrinkage data are plotted in Fig. 4-52 as a function of the non-crystalline orientation present in the sample after annealing (here the term $(1 - \beta)$ corrects for the fraction of noncrystalline chains present in the sample). The noncrystalline chains are seen to decrease in orientation with increasing temperature by an amount proportional to the shrinkage that occurred in the sample. The shrinkage mechanism in oil and air is seen to be the same, since the shrinkage–structure relation is the same for the 5 × drawn fibers in the two media, the lower shrinkage in air occurring solely as a consequence of poor heat transfer in this medium. The amount of shrinkage is also seen to be directly related to the initial structure present in the original drawn fibers before shrinkage, as the shrinkage–structure relations for the 3 × yarn and the 5 × yarn fall on two distinct lines with intercepts corresponding to the structure of the original drawn fiber in each case.

The structure produced during the initial drawing process and the subse-quent structural change that occurs on annealing are, thus, intimately

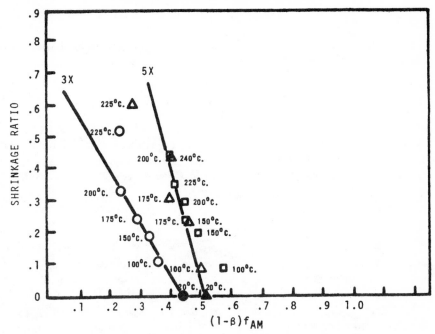

Fig. 4-52   Relation between PET orientation, crystallinity, and shrinkage. ($\triangle$) 5 × oil-annealed fibers; ($\square$) 5 × air-annealed fibers; ($\bigcirc$) 3 × oil-annealed fibers; ($\blacksquare$, $\bullet$) unannealed drawn fibers.

related. Is the character of the shrinkage process in fact determined by the fabrication conditions used to produce the original drawn fibers? If the fabrication process begins with an unoriented amorphous fiber, and it is assumed that the extension of the fiber proceeds primarily by orientation of the noncrystalline polymer, then the amount and degree of orientation of the noncrystalline polymer should be directly proportional to the extension of the fiber. If the thermally shrunk fiber is considered as the final product of a fabrication process in which the fiber was first drawn to some extension (characterized by a measurable noncrystalline polymer content and degree of orientation) and then relaxed to a lesser extension (characterized by a new noncrystalline content and degree of orientation), then there should be a direct relation between the true strain (logarithm of the extension ratio) of the final fiber and the amount and degree of orientation $[(1 - \beta)f_{am}]$ of the noncrystalline polymer in the final fiber.

The true strain is used here as the measure of extension, as this was shown in Section A on polypropylene to be the correct strain parameter to use when describing plastic deformation of a polymer. The true strain, in the polypropylene study, was found to be a linear function of the noncrystalline orientation over the whole draw range for all of the fabrication processes examined.

The original extension ratio λ describing the deformation of the un-oriented, spun fiber is given by the expression:

$$\lambda = \frac{L}{L_0} \tag{4-7}$$

where L_0 is the length of the original spun fiber and L is the length of the fiber after it has been drawn. Consider a two-stage process in which the fiber is first drawn an extension ratio λ, and subsequently relaxed (through shrinkage in air or oil). The extension ratio λ' for the two-stage process (the residual extension ratio) can be defined as:

$$\lambda' = \frac{L'}{L_0} \tag{4-8}$$

where L' is the length of the fiber at the end of the two-stage process. The residual extension ratio describes the total deformation process, while the shrinkage ratio S describes only a portion of the deformation process.

To interpret the shrinkage data from the PET fibers meaningfully, the shrinkage ratio must be converted to a residual extension ratio. The initial sample length in the shrinkage experiment (eq. 4-6) is simply the final length L from the original drawing process. The final length of the sample after shrinkage is L', the length of the fiber at the end of the two-stage process. Thus the shrinkage ratio S can be defined as:

$$S = \frac{L - L'}{L} \tag{4-9}$$

and

$$L' = L(1 - S) \tag{4-10}$$

The residual extension ratio of a fiber that has undergone a two-stage process of extension and subsequent shrinkage is, thus, given by the expression:

$$\lambda' = \frac{L'}{L_0} = \frac{L(1 - S)}{L_0} = \lambda(1 - S) \tag{4-11}$$

where λ is the original drawn ratio (extension ratio) of the PET fibers before shrinkage (in this case either $3\times$ or $5\times$) and S is the measured shrinkage ratio of the sample (see Table 4-1).

The residual extension ratio of the PET fibers is plotted as true residual strain on the ordinate of Fig. 4-53. The abscissa is $(1 - \beta)f_{am}$, which describes the state of noncrystalline orientation at the end of the fabrication

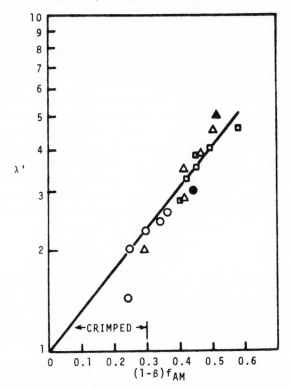

Fig. 4-53 Effect of PET orientation and crystallinity on the residual extension ratio after shrinkage: ($\triangle$) $5\times$ oil-annealed fibers; ($\square$) $5\times$ air-annealed fibers; ($\bigcirc$) $3\times$ oil-annealed fibers; ($\blacksquare$, $\bullet$) unannealed drawn fibers.

process. Included in the figure are data on the original fibers drawn $3\times$ and $5\times$, the fibers drawn $3\times$ and $5\times$ and subsequently shrunk in oil at different temperatures, and the fibers drawn $5\times$ and subsequently shrunk in air at different temperatures; a total of eighteen structural states and deformation processes. All these data fall along a single straight line with a zero intercept, suggesting that the sample orientation returns along the original extension path during shrinkage. The figure also demonstrates that all of the shrinkage behavior can be quantitatively expressed by the structural-state equation:

$$\log \lambda' = \log \lambda'_0 + K[(1 - \beta)f_{am}] \tag{4-12}$$

Here the intercept $\log \lambda'_0 = 0$ for the PET fibers and K is the slope of the line. The fact that all 18 samples fit this simple quantitative expression shows that the shrinkage mechanism is primarily controlled by the behavior of the noncrystalline chains.

(2) Crimp. Another observation made was that during oil annealing the $3 \times$ yarns developed a crimp at annealing temperatures above 175°C, while the $5 \times$ yarns did not develop crimp until an annealing temperature of 225°C. The $5 \times$ yarns did not develop crimp during air annealing.

Once one realizes that the shrinkage is controlled by the noncrystalline chains, it is reasonable to examine the question of crimp in the fiber. Crimp is characterized by a bending, curling, or crumpling of the fiber and can be considered as occurring either from a rapid "snapback" of the fiber or from some structural relaxation resulting at some minimum orientation. When the $3 \times$ drawn yarn is annealed in oil and allowed to shrink freely, crimp first occurs at 175°C. At that temperature the noncrystalline orientation has been reduced to $(1 - \beta)f_{am} = 0.297$. When the $5 \times$ drawn fiber is annealed in oil and allowed to shrink freely, crimp does not occur until the annealing temperature of 225°C is reached. At that temperature the noncrystalline orientation of the $5 \times$ sample has been reduced to $(1 - \beta)f_{am} = 0.293$. Thus in both the $3 \times$ and $5 \times$ drawn oil-shrunk samples, crimp does not appear until the noncrystalline orientation has been reduced to the same value (0.29). This seems to be a critical structural value for crimp formation in the PET fibers, a structural condition that occurs at different temperatures in the two fibers as a consequence of their initial difference in noncrystalline orientation. All fibers with $(1 - \beta)f_{am}$ value of 0.29 or less were crimped (see Fig. 4-53).

None of the thermally annealed $5 \times$ drawn samples crimped in air. The lack of crimp in these fibers is not due to some unusual air-directed shrinkage mechanism, but is simply a consequence of the poorer heat transfer in air. The poorer heat transfer led to less relaxation of the noncrystalline chains and, as a consequence, $(1 - \beta)f_{am}$ never reached the critical crimp value of 0.29. Instead, the lowest $(1 - \beta)f_{am}$ value the air-annealed sample reached was 0.408 at 240°C.

(3) Long Spacing. When the drawn PET fiber is annealed under unrestrained conditions two processes occur: (1) the length of the fiber decreases (the fiber shrinks) and (2) the x-ray small-angle long spacing increases. The small-angle long spacing is a measure of the average repeat distance of the crystal lamellae as measured parallel to the helical chain axis of the molecules. Thus the average distance between crystal centers increases as the length of the fiber decreases.

How can this seemingly anomalous result be explained? Can a structural relationship be found between the relaxation of the noncrystalline chains and the growth of the crystallites? How closely are the shrinkage and the crystallite growth related? The effect of annealing temperature and immersion medium on the long spacing of the PET fibers is shown in Fig. 4-54. After

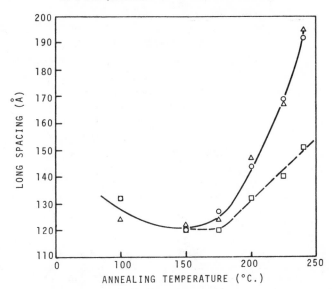

Fig. 4-54 Effect of annealing temperature and medium on the long spacing of PET fibers. (△) 5 × oil-annealed fibers; (□) 5 × air-annealed fibers; (○) 3 × oil-annealed fibers.

an initial slight decrease, the long spacing increases with increasing temperature for all of the fibers. The 3 × and 5 × drawn yarns annealed in oil follow the same long-spacing curve, increasing quite rapidly with increasing temperature. The 5 × drawn fiber annealed in air also increases in long spacing with increasing temperature, but at a much slower rate than the oil-annealed samples.

Except for the higher long spacing at an annealing temperature of 100°C, the curves for long spacing versus annealing temperature in Fig. 4-54 are very similar in shape to those of shrinkage ratio versus annealing temperature in Fig. 4-51. This suggests a direct relation between the shrinkage ratio and the long spacing developed in the PET fibers. The shrinkage ratio is plotted against the long spacing in Fig. 4-55; a linear relation is obtained for each of the PET fiber shrinkage sets, except for samples annealed at 100°C. This also suggests that the decrease in sample length and the increase in long spacing are intimately related.

The shrinkage of the fiber should be considered as the second stage of a two-stage fabrication process which includes extension as well as shrinkage. Is the size of the annealed crystal a direct function of the relaxation of the noncrystalline chains? Can the change in long spacing be directly related quantitatively to the residual extension of the two-stage fabrication process?

If the size of the crystal is controlled by the relaxation of the noncrystalline

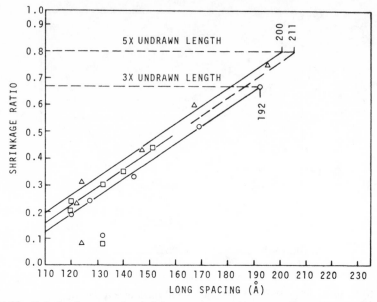

Fig. 4-55 Relation between the shrinkage ratio and the long spacing of annealed PET fibers: (△) 5× oil-annealed fibers; (□) 5× air-annealed fibers; (○) 3× oil-annealed fibers.

chains, the maximum orientation-controlled long spacing would occur when the noncrystalline orientation of the particular drawn sample is allowed to relax to the random state ($f_{am} = 0$). This condition will be achieved when the sample shrinks to its original length before extension (i.e., a residual extension ratio λ' of 1.0). For the 3 × drawn sample this occurs at a shrinkage ratio S of 0.67, while for the 5 × drawn sample this occurs at a shrinkage ratio of 0.80. If the linear relation between S and the long spacing L for each shrinkage set in Fig. 4-55 is extrapolated to the value of S corresponding to the original length of the sample, the maximum relaxation-controlled long spacing L_{max} for that set can be determined. For the 3 × drawn sample, maximum shrinkage is achieved at 240°C, and $L_{max} = 192$ Å (see Fig. 4-55). The 5 × drawn sample almost reaches L_{max} at 240°C in oil; L_{max} for the oil samples is 200 Å. The 5 × samples annealed in air are some distance from $L_{max} = 211$ Å, even at 240°C, as a consequence of poor heat transfer.

The distance of the measured long spacing L from the maximum value it could have if the fiber were allowed to shrink completely is ($L_{max} - L$). The proportion of the original extension of the fiber which this change in spacing represents is simply the product of the draw ratio λ_F and ($L_{max} - L$). This product represents the crystallite growth analog of the residual extension and normalizes all of the long-spacing data to the fabrication conditions of both drawing and shrinkage (i.e., the complete two-stage fabrication process).

In Fig. 4-56, $\lambda_F(L_{max} - L)$ is plotted against the residual extension ratio

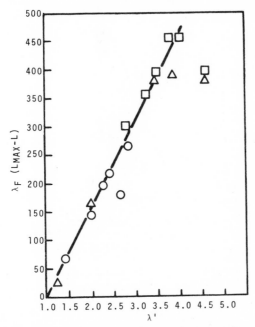

Fig. 4-56 Relation between $\lambda_F(L_{max} - L)$ and the residual extension ratio λ' of annealed PET fibers: ($\triangle$) 5 × oil-annealed fibers; ($\square$) 5 × air-annealed fibers; ($\bigcirc$) 3 × oil-annealed fibers.

λ'. All the data, except for the samples annealed at 100°C, fit a straight line with a positive slope and a zero intercept, having the form:

$$\lambda_F(L_{max} - L) = K\lambda' \qquad (4\text{-}13)$$

This includes the 3 × and 5 × drawn samples annealed in oil, and the 5 × drawn samples annealed in air. Thus, it seems, the change in long spacing with annealing of drawn fiber can be directly related quantitatively to the residual extension of the two-phase fabrication process.

The fact that the general long-spacing term is directly and quantitatively related to the residual extension suggests that the crystallization process is controlled by the relaxation of the noncrystalline chains. This conclusion is reached as a consequence of the fact that the true residual strain is directly proportional to the amount and degree of orientation of the noncrystalline

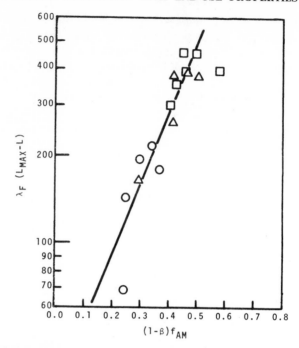

Fig. 4-57 Relation between $\lambda_F(L_{max} - L)$ and $[(1 - \beta)f_{am}]$ for annealed PET fibers: ($\triangle$) 5× oil-annealed fibers; ($\square$) 5× air-annealed fibers; ($\bigcirc$) 3× oil-annealed fibers.

chains (Fig. 4-53) and therefore $\lambda_F(L_{max} - L)$ will exhibit a similar proportionality when plotted in a similar manner (Fig. 4-57). Thus it seems that shrinkage proceeds by a two-step process in which, first, the noncrystalline chains relax along their original deformation path, causing a decrease in the length of the fibers, and then the relaxed chains are utilized by the crystals for growth.

(4) Small-Angle X-Ray (SAXS) Intensity. According to small-angle x-ray theory (35), the diffraction intensity I is proportional to the square of the difference, $\rho_c - \rho_{am}$, in electron density between the crystalline and noncrystalline regions, thus,

$$\sqrt{I} \approx \rho_c - \rho_{am} \tag{4-14}$$

For an oriented polymer the intensity is also proportional to the degree of orientation of the regions, since the discrete scattering intensity is anisotropic and only occurs parallel to the molecular helical chain axis direction (Section B.3 in Chapter 3 and Section A.1.b, this chapter).

The greatest difference in $(\rho_c - \rho_{am})$ will occur when the noncrystalline chains are fully relaxed and $f_{am} = 0$. Under this special condition the molecules in the crystalline regions are aligned and those in the noncrystalline region are disordered. This large difference in structural character will lead to the highest SAXS intensity for a given crystalline fraction. The other intensity extreme occurs when the noncrystalline region is fully oriented and all of the noncrystalline chains are aligned. For this special case the noncrystalline polymer has an ordered character more nearly approximating the ordered character of the molecules in the crystals and this structural similarity between the crystalline and noncrystalline regions leads to the lowest SAXS intensity for a given crystalline fraction. Of course, intermediate degrees of orientation of the noncrystalline chains will lead to intermediate SAXS intensity values. Thus, the small-angle x-ray diffraction intensity will be some function of the amount and degree of orientation of the polymer chains.

Since the SAXS intensity is proportional to the square of the difference in the electron density of the two regions, $\sqrt{I}$ is the proper variable to use for structural analysis. Figure 4-58 shows the effect of annealing tempera-

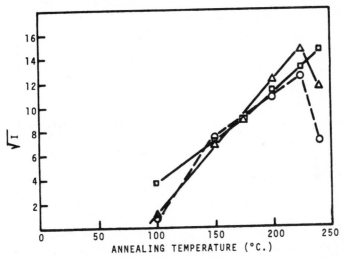

Fig. 4-58 Relation between the small-angle x-ray intensity and the annealing temperature of PET fibers. ($\triangle$) 5 × oil-annealed fibers; ($\square$) 5 × air-annealed fibers; (O) 3 × oil-annealed fibers.

ture and medium on the observed SAXS intensity. Except for the PET fibers annealed in oil at 240°C, for which there are no structural data, the SAXS intensity increases with increasing annealing temperature. This character of the annealing curve is similar to that previously observed for the effect of annealing temperature on the shrinkage ratio (Fig. 4-51) and the long spacing (Fig. 4-54).

In view of the similarity of the SAXS behavior to that of the shrinkage ratio S and the long spacing L, and the fact that SAXS intensity, like S and L, will depend on the orientation of the polymer, the SAXS intensity was plotted against the noncrystalline orientation $(1 - \beta)f_{am}$ (Fig. 4-59). The

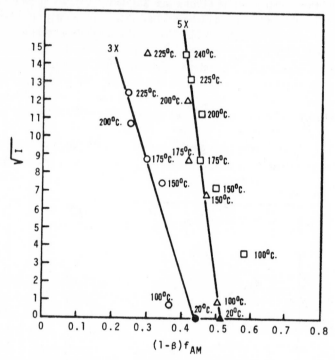

Fig. 4-59 Relation between the small-angle x-ray intensity and the noncrystalline orientation of annealed PET fibers: ($\triangle$) $5\times$ oil-annealed fibers; ($\square$) $5\times$ air-annealed fibers; ($\bigcirc$) $3\times$ oil-annealed fibers; ($\bullet$, $\blacktriangle$) unannealed drawn fibers.

dependence of the SAXS intensity on the noncrystalline orientation is obvious, and almost identical with that previously observed for the shrinkage ratio (Fig. 4-52). Thus, the SAXS intensity is seen to increase with increasing temperature as the noncrystalline orientation decreases and the sample shrinks. The character of the SAXS intensity increase is the same whether it occurs in an oil or an air medium, since the relation between $\sqrt{I}$ and structure is the same for the $5\times$ drawn fibers annealed in both media. This suggests that the same noncrystalline chain disorientation mechanism is responsible for the intensity increase in both media, a conclusion identical with that for the shrinkage process. The increase in $\sqrt{I}$ is also seen to be directly related to the initial starting structure present in the drawn fibers before shrinkage, as the $\sqrt{I}$–structure relations for the 3 × yarn and the 5 × yarn fall on two distinct lines with intercepts corresponding to the structure of the original drawn fibers.

The similarity between the SAXS intensity behavior and the behavior of the shrinkage ratio suggests that there is a direct relationship between these two parameters. The shrinkage ratio is plotted against the SAXS intensity $\sqrt{I}$ in Fig. 4-60. All the data, except for the PET fibers annealed in oil at

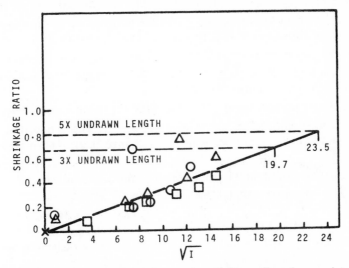

Fig. 4-60 Relation between the shrinkage ratio and the small-angle x-ray intensity of annealed PET fibers: (△) 5 × oil-annealed fibers; (□) 5 × air-annealed fibers; (○) 3 × oil-annealed fibers; (×) unannealed drawn fibers.

240°C, for which there are no structural data and which behave anomalously throughout (see Fig. 4-58), follow a single straight line with zero intercept. The experimental magnitude of $\sqrt{I}$ for the unannealed drawn fibers is taken as zero, as the integrated intensity of these samples was reported as "so low as to be hardly measurable for the 3 × and 5 × samples as drawn, and there appeared to be little intensity difference between the samples" (32). The fact that within the scatter of the data there is a direct relation between the shrinkage ratio and the $\sqrt{I}$, shows that the decrease in sample length and the increase in SAXS intensity are intimately related.

The shrinkage of the fiber must be considered as the second stage of a two-stage process. The SAXS intensities from both the drawing process and from the subsequent shrinkage process are intimately related (Fig. 4-59). The SAXS data must be treated in terms of the total two-stage process before a single quantitative structural correlation can be achieved. The maximum anisotropic structure-controlled SAXS intensity will occur when the noncrystalline chains are allowed to relax to the random state ($f_{am} = 0$) under those thermal conditions that achieve the largest, most perfect crystals. This will occur when the drawn sample is allowed to shrink freely back to the undrawn length of the original spun-fiber residual extension ratio ($\lambda' = 1.0$). As was pointed out in the discussion of the long spacing, this occurs at a shrinkage ratio of 0.67 for the 3 × drawn fiber, and at a shrinkage ratio of 0.80 for the 5 × drawn fiber. By extrapolating the linear relation between the shrinkage ratio and $\sqrt{I}$ to the corresponding shrinkage ratio for complete shrinkage of that fiber ($\lambda' = 1.0$), the value of the maximum anisotropic structure-directed SAXS intensity $\sqrt{I_{max}}$ can be determined. Values thus obtained are $\sqrt{I_{max}} = 19.7$ for the 3 × drawn fiber, and $\sqrt{I_{max}} = 23.5$ for the 5 × drawn fiber (Fig. 4-60).

The amount the measured $\sqrt{I}$ differs from the value it could have if full shrinkage were allowed to occur is $\sqrt{I_{max}} - \sqrt{I}$. The proportion of the original extension of the fiber which this change in spacing represents is simply $\lambda_F(\sqrt{I_{max}} - \sqrt{I})$ which represents the SAXS intensity analog of the residual extension, and normalizes all of the SAXS intensity data to the conditions of the two stage fabrication process.

In Fig. 4-61, $\lambda_F(\sqrt{I_{max}} - \sqrt{I})$ is plotted against the residual extension ratio λ'. All of the data, except for the sample annealed in oil at 240°C, fit a straight line with positive slope and zero intercept, having the form,

$$\lambda_F(\sqrt{I_{max}} - \sqrt{I}) = A\lambda' \qquad (4\text{-}15)$$

Thus it seems the change in SAXS intensity that occurs when drawn PET fibers are thermally annealed and allowed to shrink freely is directly related quantitatively to the residual extension of the two-stage fabrication process.

The SAXS intensity will increase as the oriented noncrystalline chains relax, disorient, and become less crystallike. Since $\lambda_F(\sqrt{I_{max}} - \sqrt{I})$ is directly proportional to the residual true strain, which is in turn directly proportional to the orientation of the noncrystalline chains (Fig. 4-53), the correlation between the SAXS intensity and the noncrystalline orientation can be placed on a quantitative basis. A plot of $\log[\lambda_F(\sqrt{I_{max}} - \sqrt{I})]$ against $(1 - \beta)f_{am}$ is linear (Fig. 4-62), and can be expressed as,

$$\log[\lambda_F(\sqrt{I_{max}} - \sqrt{I})] = A + B[(1 - \beta)f_{am}] \qquad (4\text{-}16)$$

where A and B are constants. This expression fits data from the drawn and unannealed fibers, as well as the data from the drawn fibers shrunk in different media at different annealing temperatures, demonstrating that SAXS intensity data can be quantitatively correlated with the structural state of the polymer.

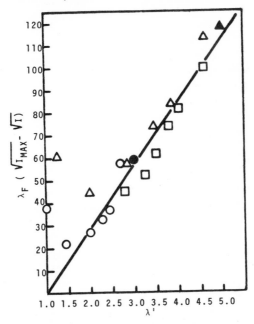

Fig. 4-61 Relation between $\lambda_F(\sqrt{I_{max}} - \sqrt{I})$ and the residual extension ratio λ' of annealed PET fibers: ($\triangle$) 5× oil-annealed fibers; ($\square$) 5× air-annealed fibers; ($\bigcirc$) 3× oil-annealed fibers; ($\bullet$, $\blacktriangle$) unannealed drawn fibers.

The thermal annealing of the PET fibers is part of a total fabrication process whereby the fiber is drawn from one structural state (that of the spun fiber), to a second structural state (that of the drawn fiber), and then to a final structural state (that of the fiber after shrinkage). The change in the properties of the sample represents the change that has occurred in the fiber as it goes from its initial state to the final structural state, and is intimately related to the state functions. In order to show how the structural-state change has produced the observed changed properties of the polymer it is necessary to describe the total change of the property in terms of the initial and final stage of the fabrication process. Thus, shrinkage ratio must be converted to the residual extension; long spacing must be related to the draw ratio and the final recovered spacing; and SAXS intensity must be treated similarly. Once this is done, the intimate relation among shrinkage, crimp, long spacing, and SAXS intensity, and their quantitative identification with the change in the noncrystalline orientation states becomes obvious.

In this way, thermal annealing can be understood mechanistically as a process by which a sample that has been drawn to some orientation state, gets enough thermal energy from its environment to allow the noncrystalline

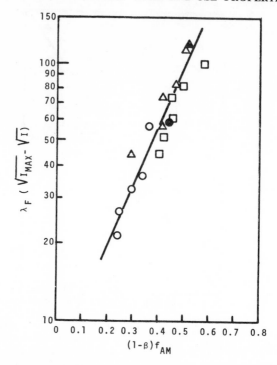

Fig. 4-62 Relation between $\lambda_F(\sqrt{I_{max}} - \sqrt{I})$ and $(1 - \beta)f_{am}$ for annealed PET fibers: ($\triangle$) $5\times$ oil-annealed fibers; ($\square$) $5\times$ air-annealed fibers; ($\bigcirc$) $3\times$ oil-annealed fibers; ($\bullet$, $\blacktriangle$) unannealed drawn fibers.

chains to disorient. The decrease in the chain alignment causes a decrease in the length of the fiber (shrinkage) and makes less strained chains available for the crystals to use for growth (long spacing increases). The increasing difference between the ordered molecular arrangement within the annealed crystals, and the disordered arrangement of the noncrystalline chains as the temperature is raised, leads to an increase in the SAXS intensity with increasing annealing temperature. If the disordering of the noncrystalline chains goes beyond a critical value, the sample crimps. Above that critical value no crimp will appear. Thus the property changes that occur on thermal annealing are all intimately and quantitatively correlated to the changing structural state of the polymer.

c. Dynamic Loss Modulus. The dynamic mechanical loss modulus is a measure of the energy dissipated by the polymer as a consequence of molecular relaxation processes. When the loss modulus is measured at a fixed frequency at different temperatures a dynamic spectrum will be produced.

At specified temperatures in the spectra the loss modulus will have maxima, which represent specific molecular relaxation processes. In PET the α loss peak, E''_{max}, represents energy dissipated by molecular relaxation processes in the noncrystalline region of the polymer (36). It is, in effect, a measure of the glass transition temperature T_g of the polymer.

The molecular mechanism that occurs in the thermal relaxation process involves a transition from a thermally frozen molecular (glassy) state to one in which the molecules are allowed to relieve the frozen-in stresses by movement in the noncrystalline region. Though the ends of the sample are restrained during the dynamic loss modulus measurement, the molecular motions that occur at the α transition are similar in kind to those that occur from thermal relaxation of the built-in stresses during free shrink. The loss modulus maximum E''_{max} will be reached only when the temperature has imparted enough energy to the molecules to overcome the thermodynamic stability of the particular state represented by the orientation of the molecules in the noncrystalline region of the polymer. Shrinkage is controlled by similar forces. Thus, just as with shrinkage, the observed temperature at which the α process, E''_{max}, occurs in the PET fibers should be some function of the amount and degree of orientation of the noncrystalline chains in the starting sample.

The starting fibers used for the loss modulus measurements had a complicated fabrication history (31). PET yarns initially drawn $3 \times$ and $4.25 \times$ over a hot pin at 80°C were subsequently heated for 6 hr in vacuo at temperatures up to 240°C and allowed to shrink freely. These shrunk fibers were then given a further treatment in which they were boiled in water, under relaxed conditions, for 1 hr. The structural state present in the fibers after this complicated fabrication procedure (Table 4-2) is the one that determines the subsequent loss modulus results.

The relation between T_{max}, the measured temperature of E''_{max} for the final fabricated fibers, and the annealing temperature used during the initial shrinkage in vacuo is shown in Fig. 4-63. The essentially flat region of the curves for annealing temperatures from room temperature to 125°C occurs as a consequence of the subsequent boiling-water fabrication step, in which the fibers annealed at lower temperature are all further annealed to a structural state characteristic of the higher boiling-water temperature. As the other fibers were initially relaxed during free shrinkage at temperatures higher than that of boiling water, their final structure was predominantly developed during the initial shrinkage in vacuo. Both the $3 \times$ and $4.25 \times$ drawn and annealed fibers show a continual decrease in T_{max} in Fig. 4-63 with increasing temperature of thermal annealing after 125°C.

The manner in which the PET α loss peak maximum temperature T_{max} depends on the initial structural state of the test fibers (Table 4-2) is shown in Fig. 4-64. It is evident that T_{max} is directly proportional to $(1 - \beta)f_{am}$ (the

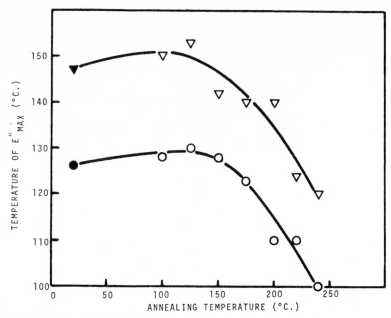

Fig. 4-63 Relation between the temperature of E''_{max} and the annealing temperature of PET fibers: (○) 3× annealed fibers; (▽) 4.25× annealed fibers; (●,▼) unannealed drawn fibers.

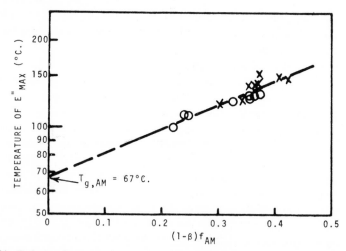

Fig. 4-64 Relation between the temperature of E''_{max} and the amount and degree of orientation of the noncrystalline polymer: (○) initial draw 3×; (×) initial draw 4.25×.

structural state of the noncrystalline region) for all of the 16 structural states examined. When the best line through the points is extrapolated to the fully amorphous, unoriented state $[(1 - \beta)f_{am} = 0]$, a T_{max} of 67°C is predicted. The known glass transition temperature T_g of unoriented fully amorphous PET is 67°C (37).

The fact that the temperature of E''_{max} depends on both the amount and state of orientation of the noncrystalline polymer is not unexpected. The effect of orientation on the molecular mobility of crystallizable polystyrene (10), nylon 66 (11–13) and PET (14,15) has been examined by nuclear magnetic resonance, dynamic mechanical, and thermal distortion techniques. These studies all showed that there is an increase in resistance to thermal mobility of the molecules with increasing orientation, and the higher the molecular orientation, the higher is the observed T_g. What these studies did not show, as the orientation state of the test samples had not been quantitatively defined, was any quantitative correlation between the orientation state and the thermal mobility of the molecules. Figure 4-64 shows such a quantitative correlation:

$$\log T_{max} = \log T_{g,am} + K(1 - \beta)f_{am} \qquad (4\text{-}17)$$

where K is the slope constant. Thus, T_{max}, just as the tenacity, shrinkage, long spacing, and SAXS intensity, can be characterized in terms of structural mechanisms, provided the structural state of the polymer is quantitatively defined.

d. Tensile Modulus.   The tensile modulus of the fiber should be a measure of the slope of the stress–strain curve in the region of the elastic limit (Fig. 4-48). Since for polycrystalline polymers, the elastic limit occurs at very low extensions ($<0.5\%$), the observed stress–strain curve, even in its observed lower limit, only approximates elastic behavior. For this reason the measured tensile modulus usually includes some viscoelastic contributions, especially at the slow rates of strain generally used for tensile measurements.

When a high-frequency sound pulse is sent along a fiber, the sound wave propagates by molecular compression and extension (i.e., a stress–strain curve of molecular dimensions). The combination of high frequency (in the kilohertz range) and small displacement of the polymer by the sound wave, makes the sample respond elastically. As no structural changes occur from pulse propagation, the observed sonic velocity depends on the structure of the polymer investigated. Since the sonic modulus E_s is simply the density times the square of the sonic velocity, the modulus also depends directly on the average structure of the polymer investigated.

The relation between the sonic modulus and the internal structure of a uniaxially oriented polymer is given by the expression (see Chapter 2,

Section B.3):

$$\left(\frac{3}{2}\right)\Delta E^{-1} = \frac{\beta f_c}{E_{t,c}^0} + \frac{(1-\beta)f_{am}}{E_{t,am}^0} \qquad (2\text{-}35)$$

where

$$\Delta E^{-1} = E_u^{-1} - E_{or}^{-1}$$

and

$$E_u^{-1} = \left(\frac{2}{3}\right)\left(\frac{\beta}{E_{t,c}^0} + \frac{1-\beta}{E_{t,am}^0}\right) \qquad (2\text{-}34)$$

Here, E_{or} is the sonic modulus of an oriented sample; E_u is the sonic modulus of an unoriented sample of the same crystallinity as the oriented sample; $E_{t,c}^0$ and $E_{t,am}^0$ are the intrinsic lateral Young's moduli of the crystalline and noncrystalline regions, respectively; and β, f_c, and f_{am} have definitions given previously.

The intrinsic lateral moduli of the crystalline and noncrystalline regions of PET are 3.68×10^{10} dyne/cm^2 and 1.82×10^{10} dyne/cm^2, respectively [4]. Converting this information to grams per denier gives

$$E_s\left(\frac{g}{denier}\right) = 0.1133 \times 10^{-8}\, \frac{E_s(\text{dyne/cm}^2)}{\text{density (g/cm}^3)} \qquad (4\text{-}18)$$

and applying this, together with the structure data in Table 4-1, to eq. 2-35 yields E_s(g/denier) for the oriented PET fibers.

The measured tensile moduli E_{TM} and structure-calculated sonic moduli E_s of the thermally annealed PET fibers are not expected to be identical for the same sample, as a consequence of the extreme strain-rate difference between the two measurements, and the fact that some viscoelastic mechanisms influence the tensile modulus. Their trends should be similar, however, if the tensile modulus is structure dominated.

The similar structure dependence of the measured tensile modulus and the predicted sonic modulus (calculated from structure data) can be readily demonstrated for the PET fibers. Equation 2-35 can be rearranged to yield:

$$\frac{1}{E_{or}} = \left[\left(\frac{1}{E_u}\right) - \left(\frac{2\beta f_c}{3E_{t,c}^0}\right)\right] - \left(\frac{2}{3E_{t,am}^0}\right)[(1-\beta)f_{am}] \qquad (4\text{-}19)$$

Since the crystallinity and crystallite orientation data in Table 4-1 are fairly constant, the bracketed term $[(1/E_u) - 2\beta f_c/3E_{t,c}^0]$ can be treated as a constant (except for four samples, the bracketed term has values between 0.027 and 0.029 with an average value of 0.0277). Thus eq. 4-19 predicts, specifically for the data in Table 4-1, that a linear relation exists between the calculated sonic compliance $(1/E_{or})$ of the oriented fibers and $(1-\beta)f_{am}$. Figure 4-65 shows a plot for all of the samples in Table 4-1. The straight line was drawn

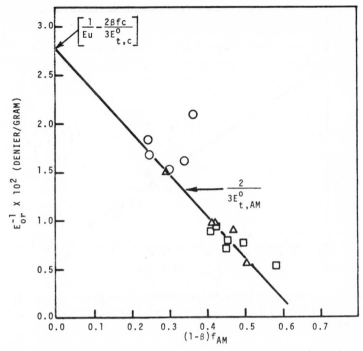

Fig. 4-65 Theoretical correlation between the sonic compliance E_{or}^{-1} (calculated from β, f_c, and f_{am}) and $(1 - \beta)f_{am}$: ($\triangle$) 5× oil-annealed fibers; ($\square$) 5× air-annealed fibers; (O) 3× oil-annealed fibers.

by using the average-intercept value (0.0277) calculated from the structure data in Table 4-1 and the known intrinsic sonic moduli, and a slope calculated from the known value of $E_{t,am}^0$. All the data fit the predicted line except for the 3× drawn yarn shrunk at 100°C. This sample has a different crystallinity than that of the other samples and therefore does not fit the constant βf_c criterion of eq. 4-19.

If the measured tensile modulus has a structure dependence similar to that of the calculated sonic modulus, it should be revealed in a plot of the tensile compliance $(1/E_{TM})$ against $(1 - \beta)f_{am}$. This is because the tensile modulus data are also for the samples in Table 4-1, where the crystallinity and crystallite orientation are relatively constant. Figure 4-66 shows a plot of the measured tensile compliance $(1/E_{TM})$ against $(1 - \beta)f_{am}$ for the samples in Table 4-1. Except for the 3× and 5× samples that crimped in oil at 225°C, all of the data fit the predicted relation quite satisfactorily. Of course, the intercept and slope are no longer the structure-predicted parameters, since the tensile modulus data are taken at a slow speed and have viscoelastic components. The similarity between the predicted structure

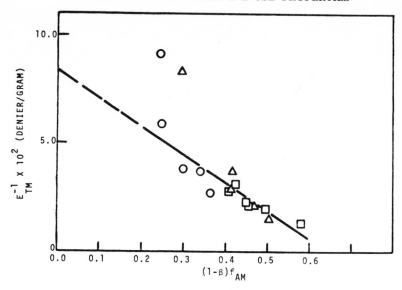

Fig. 4-66 Experimental correlation between the tensile compliance E_{TM}^{-1} and $(1 - \beta)f_{am}$ for PET fibers: ($\triangle$) 5× oil-annealed fibers; ($\square$) 5× air-annealed fibers; ($\bigcirc$) 3× oil-annealed fibers.

dependence of the sonic modulus and the observed structure dependence of the tensile modulus is striking. Thus, the measured tensile modulus behaves as predicted by theory, and like the other properties of PET is dominated by the fabricated structural state of the samples used in the measurement.

3. Summary

The purpose of this section was to show how the same quantitative morphological criteria used to define the structural state of isotactic polypropylene in Section A leads to similar quantitative structure–property correlations when applied to PET. The data examined included tenacity, thermal shrinkage in oil and air, tensile modulus, dynamic loss modulus, long spacing, and SAXS intensity data, along with structural data for quantitatively describing the structural state of the sample.

When the tenacity, shrinkage, and tensile modulus were examined in terms of defined structural criteria, the quantitative PET structure–property correlations had the form predicted from the study of isotactic polypropylene. The tenacity correlated with amorphous orientation; shrinkage correlated directly with amorphous orientation provided, as required by the earlier study, it was treated as a true residual extension; and the tensile modulus followed the form required by the structural-state–sonic-modulus equations, all exactly as predicted from the work on isotactic polypropylene. Since there were no discrepancies between the predicted and observed behavior of the PET, it can be concluded that the structural-state approach to physical-property correlations is a general one not specific to isotactic polypropylene.

Structure–property correlations for the dynamic loss modulus, long spacing, and SAXS intensity had not been examined in the study of isotactic polypropylene. The general validity of the structural-state approach to structure–property correlations is, therefore, further reinforced by the quantitative correlations observed between these properties and the structural state of PET fibers. Thus, the temperature at which the maximum in the α loss modulus appears is quantitatively correlated with the amount and degree of orientation in the test fibers. Also, thermal annealing with free shrinkage is shown to be a two-stage process whereby the fiber shrinks by relaxation of the noncrystalline chains, which then become available for incorporation into the annealing crystallite. This dependence of crystal growth on orientation relaxation leads to quantitative correlation between the increase in long spacing and the disorientation of the noncrystalline chains. Finally, an increase in SAXS intensity with increased thermal annealing temperature in the free-shrink experiments is shown to be due to the disorientation of the noncrystalline chains. A quantitative correlation is obtained between the SAXS intensity and the orientation state of the non-crystalline chains in the PET fibers.

In all of the cases studied, examination of the properties of the polymer in terms of the structural state of the sample has led to simplified quantitative correlations. The internal mechanisms controlling the observed property became obvious, and served as a guide to further correlations. Fabrication parameters such as draw ratios, rates of draw, and draw or annealing temperature, though interesting in their historical context, could not be used as generalizing criteria for property analysis. Only by considering the internal structure produced by the particular fabrication process could this be done. Thus the major conclusion of this section must be that only by considering the structural state of a polymer can general, simplifying, quantitative, structure–property correlations be achieved.

References

1. R. J. Samuels, *J. Polymer Sci.*, *A*, **3**, 1741 (1965).
2. R. J. Samuels, *J. Polymer Sci.*, (*A-2*), **6**, 1101 (1968).
3. R. J. Samuels, *J. Polymer Sci.*, (*A-2*), **7**, 1197 (1969).
4. J. H. Dumbleton, *J. Polymer Sci.*, (*A-2*), **6**, 795 (1968).
5. R. J. Samuels, in *Science and Technology of Polymer Films*, O. J. Sweeting, Ed., Wiley–Interscience, New York, 1968, Chap. 7.

6. W. Hayden, W. G. Moffatt, and J. Wulff, *Mechanical Behavior*, Wiley, New York, 1965, Chap. 1.

7. R. J. Samuels, in *Supramolecular Structure in Polymers*, P. H. Lindenmeyer, Ed., Wiley–Interscience, New York, 1967, p. 253.

8. R. J. Samuels, *J. Polymer Sci.*, (*A-2*), **6**, 2021 (1968).

9. F. J. Balta-Calleja and A. Peterlin, *J. Polymer Sci.*, (*A-2*), **7**, 1275 (1969).

10. S. Newman and W. P. Cox, *J. Polymer Sci.*, **46**, 29 (1960).

11. A. M. Thomas, *Nature*, **179**, 862 (1957).

12. P. E. McMahon, *J. Polymer Sci.*, *B*, **4**, 43 (1966).

13. W. O. Statton, in *Supramolecular Structure in Polymers*, P. H. Lindenmeyer, Ed., Wiley–Interscience, New York, 1967, p. 117.

14. A. R. Thompson and D. W. Woods, *Trans. Faraday Soc.*, **52**, 1383 (1956).

15. I. M. Ward, *Textile Res. J.*, **31**, 650 (1961).

16. A. V. Tobolsky, *Properties and Structure of Polymers*, Wiley, New York, 1960.

17. A. M. Freudenthal, in *Fracture*, H. Liebowitz, Ed., Vol. 2, Academic, New York, 1968, Chap. 6.

18. R. F. Boyer, Ed., *Transitions and Relaxations in Polymers*, Wiley–Interscience, New York, 1966.

19. R. F. Landel and R. F. Fedors, *Fracture Processes in Polymeric Solids*, B. Rosen, Ed., Wiley–Interscience, New York, 1964, p. 361.

20. S. G. Turley and H. Keskkula, in *Transitions and Relaxations in Polymers*, R. F. Boyer, Ed., Wiley–Interscience, New York, 1966, p. 79.

21. K. M. Sinnott, *SPE Trans.*, **2**, 65 (1962).

22. W. P. Slichter and E. R. Mandell, *J. Appl. Phys.*, **29**, 1438 (1958).

23. J. A. Sauer, R. A. Wall, N. Fuschillo, and A. E. Woodward, *J. Appl. Phys.*, **29**, 1385 (1958).

24. A. E. Woodward, in *Transitions and Relaxations in Polymers*, R. F. Boyer, Ed., Wiley–Interscience, New York, 1966, p. 92.

25. W. C. Sheehan and T. B. Cole, *J. Appl. Polymer Sci.*, **8**, 2359 (1964).

26. J. A. Sauer, G. C. Richardson, and D. R. Morrow, *J. Macromol. Chem.–Rev. Macromol. Chem.*, **C9(2)**, 149 (1973).

27. P. J. Holdsworth and A. Keller, *J. Polymer Sci.*, (*A-2*), **6**, 707 (1968).

28. P. I. Vincent, *Plastics* (*London*), **26**, 141 (1961).

29. L. Ongchin and S. Sternstein, *Bull. Am. Phys. Soc.*, **14**, 362 (1969).

30. R. J. Samuels, *J. Macromol. Sci.-Phys.*, *B*, **4**, 701 (1970).

31. J. H. Dumbleton, T. Murayama, and J. P. Bell, *Kolloid Z. Z. Polym.*, **228**, 54 (1968).

32. J. H. Dumbleton, *J. Polymer Sci.*, *A-2*, **7**, 667 (1969).

33. J. H. Dumbleton, *Polymer*, **10**, 539 (1969).

34. E. I. Du Pont, Brit. Pat., 1,006,136 (1962).

35. B. K. Vainshtein, *Diffraction of X-Rays by Chain Molecules*, Elsevier, New York, 1966, p. 383.

36. N. G. McCrum, B. E. Read, and G. Williams, *Anelastic and Dielectric Effects in Polymer Solids*, Wiley, New York, 1967, pp. 501–520.

37. J. Brandrup and E. H. Immergut, Eds., *Polymer Handbook*, Interscience, New York, 1967, p. VI-87.

5

Concluding Remarks

For any area of investigation to be considered successful it should have practical as well as theoretical utility. Thus, one important aspect of structural studies should be the application of structural information to the solution of practical polymer problems. Until recently, there has been no bridge connecting the observed structural data to practical fabrication and test information. The development of the concept of equivalent structural states, in conjunction with the identification of the required structural parameters for identifying those states, has supplied that bridge. It is now possible quantitatively to analyze fabrication and test procedures structurally, and to identify which structural elements control the observed phenomena. Chapters 3 and 4 have amply demonstrated the validity of this conclusion.

It is important to recognize the practical potential of structural information, and the vast range of problem areas to which it applies. Already, in Chapter 4, the predictive capacity of structure–property correlations can be observed. Thus, it is now obvious that one must produce a high noncrystalline orientation in order to obtain a high tenacity in a crystalline polymer; that yield strength depends on the average orientation in isotactic polypropylene; that sample modulus depends on the proper balance of crystallinity, crystalline orientation, and noncrystalline orientation; that maximum resilience depends on achieving a highly oriented spherulitic structure; and so forth. All of these observations predict the structural requirements necessary for achieving the desired end-use properties.

This structure–property information is not academic. For example, all pilot plant (and operating plant) equipment is limited in the temperature and rate ranges available to them. Thus, the choice of a particular piece of equipment or a particular process approach for the evaluation of candidate materials may, without adequate structural information, result in elimination of the more superior material. For example, if a high-modulus and high-tenacity material is required, it must have a high noncrystalline orientation as well as a high crystallinity. This can require temperature, rate, and draw capabilities for one polymer lot that are outside the capacity of the pilot equipment, while these same equipment capabilities may be within the useful range for a second polymer lot. The polymer lot with superior modulus and tenacity capabilities may not be the one processible by the chosen fabrication equipment, however. If the fabrication process is not monitored structurally this instrument-fabrication limitation will not be recognized, and the best material will be discarded as a poor candidate.

The above example introduces the question of evaluation efficiency and development cost. With the present availability of techniques for evaluating the quantitative relations that exist between a polymer's structure and its properties, it should no longer be necessary or desirable initially to evaluate the suitability of a new material as a potential product by the expensive route of pilot plant studies. Instead, only a small amount of material would be required to evaluate its ultimate properties. Samples could be prepared which are *structurally known* to cover a wide range of orientations and crystallinities by commercially inefficient, but practically efficient, methods for small samples (e.g., orienting on a stretching stand, oven annealing, creep orientation using different weights, etc.). Property evaluation of this small but *structurally identified* number of samples would quickly and quantitatively establish the polymer's structure–property capacity. Armed with this knowledge an intelligent decision could be made of the product potential of the candidate polymer. Once the desired internal structures needed for particular properties have been identified, a sensible process for achieving the *cheapest* route to the required structure can be considered. Since the desired structure would be identified in advance, structural monitoring of the pilot plant studies could also minimize the expensive trial and error character of fabrication processes that is so often present for lack of suitable alternatives.

Of course, since polymer physics is an area of investigation undergoing rapid progress, new applications for structural information are constantly appearing. Thus, the author has recently extended his work on isotactic polypropylene along several directions. For example, by examining one other series of drawn films, samples drawn at 127°C, and combining the strain data for this series (obtained directly from the fabrication equipment) with that from the Series A, B, and D drawn films and fibers, it has been possible to obtain the activation energy of the noncrystalline loss processes that occurred during fabrication. Thus it was possible to get quantitative thermodynamic information by evaluating the relation between on-line process variables and the resulting internal structures produced by the process. In this way it was found that thermally activated viscous loss processes were controlling the orientability of the noncrystalline chains in the isotactic polypropylene cases studied (1).

Guided by the shrinkage information from Chapter 4, Section B, it has been possible to determine the controlling mechanism in the shrinkage of isotactic polypropylene films (1). Interestingly, the same relaxation of orientation of the noncrystalline chains that controls shrinkage in PET was found to control shrinkage in isotactic polypropylene. Since PET is 60–70% noncrystalline, while isotactic polypropylene is only 30–40% noncrystalline, this suggests the shrinkage mechanism is general for polycrystalline polymers. In fact, since many of the quantitative structure–property correlations examined seem to be valid for polycrystalline polymers in general, it should now be possible to study how chemical substituents on the polymer chain affect the properties of the polymer. This could be very helpful in developing new chemical structures that have property combinations never before achieved in a laboratory; for synthetic polymers still have a long way to go before they can duplicate the best in nature. For example, the drag-line of a spider combines the tenacity of high-tenacity nylon with the extensibility of orlon, a most desirable combination of properties. To make matters worse, the spider spins this unique fiber at ambient temperature from aqueous solution.

This brings up the subject of natural polymers. Many natural polymers are spherulitic and these polymers are structurally characterizable at all structural levels by the techniques described here. A large number of natural polymers have a more rodlike, rather than spherulitic character, however, which makes them seem less amenable to structural characterization. Such polymers as cellulose, collagen (skin, bone, eye cornea, etc.), and DNA fall into this class of materials. They are, however, just as amenable to structural analysis as are the spherulitic polymers. The same techniques apply, and only the small-angle light scattering need be of concern. The author has comprehensively evaluated a rodlike cellulosic polymer (hydroxypropylcellulose) (2), and found the same two-phase model and structural tools used for spherulitic polymers apply here. Also the light-scattering theory for SALS from rods has been developed (3,4), and offers no analytical impediment to the investigation. Since both the analytical approach and many of the structural tools are of recent origin, entirely new and important structural insights can be expected to emerge from the study of natural polymers by these techniques.

The techniques and principles developed here are also applicable to heterophase noncrystalline systems. Thus infrared dichroism is an excellent tool for evaluating the orientation of individual phases in a heterophase block copolymer system. Small-angle x-ray, sonic modulus, and birefringence, will also be useful for examination of these materials. Mechanically mixed heterophases can also be examined in this manner; stress-crystallizing rubbers may be examined fruitfully by these techniques; and there is reason to believe useful application could be made of these techniques in the examination of filled rubbers.

Table 5-1. Experimental and Theoretical Values of the Longitudinal Sonic Modulus for Aluminum/Magnesium Composites

Fraction of Metal		Sonic Modulus $(dyne/cm^2) \times 10^{-3}$		
Aluminum β	Magnesium $(1 - \beta)$	Experimental (5)	Composite Theory (6)	Samuels' Theory (7)
1.00	0.00	10.00	10.00	10.00
0.90	0.10	9.13	9.50	9.08
0.75	0.25	8.15	8.75	7.98
0.60	0.40	7.16	8.00	7.14
0.40	0.60	6.18	7.00	6.25
0.10	0.90	5.30	5.50	5.26
0.00	1.00	5.00	5.00	5.00

There is even a possibility some of the concepts developed here may be applicable to mixed metal systems. A study of the sonic properties of the mixed-metal system, aluminum/magnesium has been reported (5). The value of the sonic modulus obtained from longitudinal wave-propagation measurements is reported as a function of Al/Mg composition. The authors report they know of no theoretical or experimental values with which to compare their results. The data from their Fig. 7 are given in Table 5-1 and shown

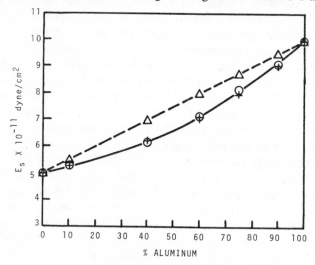

Fig. 5-1 Comparison of the theoretical and experimental values of the longitudinal sonic modulus in aluminum/magnesium composites: (○) Experimental (5); (△) Composite theory (6); (+) Samuels' theory (7); Eq. 2-34.

in Fig. 5-1. Also shown in the table and figure are the calculated modulus values for the metal composites using the longitudinal modulus equation for unidirectional composites from composite theory (6). The composite equation has the form:

$$E_L = V_F E_F + V_m E_m$$

where E_L is the longitudinal modulus, V_F is the volume fraction of fiber, E_F is the modulus of the pure fiber, V_m is the volume fraction of the matrix, and E_m is the modulus of the pure matrix. This composite equation does not satisfy the observed sonic modulus results.

Another approach is simply to use eq. 2-34 to calculate the sonic modulus of an unoriented two-phase system (7). Substituting the modulus values reported for pure magnesium and for pure aluminum into eq. 2-34 yields $E_{t,Mg}^0 = 3.33 \times 10^{11}$ dyne/cm^2 and $E_{t,Al}^0 = 6.67 \times 10^{11}$ dyne/cm^2. With the intrinsic sonic moduli now known, the sonic modulus of the Al/Mg mixtures can be calculated from the equation,

$$\frac{3}{2E_s} = \frac{\beta}{E_{t,Al}^0} + \frac{1 - \beta}{E_{t,Mg}^0} \tag{5-1}$$

The results of this calculation are shown in Table 5-1 and Fig. 5-1. The fit between the theoretical and experimental values is excellent. It is interesting to note that the transverse modulus composite equation for a unidirectional composite (6) is almost identical to eq. 2-34. Thus, use of the transverse modulus composite equation instead of the longitudinal modulus composite equation to analyze the longitudinal modulus of the mixed metal system will lead to the same numerical results as obtained with eq. 2-34.

The range of systems that are amenable to study by these structural techniques is thus quite vast. Structural examination of any one of these systems should lead to new insights and information about the structural dependence of observed properties. Such information should have definite practical applications as it reveals unexpected correlations and new ways to consider a material.

Since structural polymer physics is in its infancy, much work needs to be done. Also, as only a small fraction of the limitless areas of interest have been explored, a great deal remains to be learned. For these reasons, all who decide to embark on this search can be guaranteed an exciting, often surprising, and always very satisfying adventure.

References

1. R. J. Samuels, *J. Macromolecular Science-Phys.*, **B8(1)**, 41 (1973).
2. R. J. Samuels, *J. Polymer Sci.*, A-2, **7**, 1197 (1969).

3. M. B. Rhodes and R. S. Stein, *J. Polymer Sci.*, *A-2*, **7**, 1539 (1969).
4. R. J. Samuels, *Small-Angle Light Scattering from Polymers*, in press.
5. A. E. Lord and D. R. Hay, *J. Composite Materials*, **6**, 278 (1972).
6. R. L. McCullough, *Concepts of Fiber–Resin Composites*, Dekker, New York, 1971, p. 32.
7. R. J. Samuels, *J. Polymer Sci.*, *A*, **3**, 1741 (1965).

Index

a-Axis orientation, isotac-
    tic polypropylene,
    39-41, 116-121, 138,
    139, 142-145
Amorphous regions, 3, 17
Amorphous orientation, 20,
    41-42, 50-51, 57-63,
    119-121, 148, 154,
    157, 208, 213-241,
    243

Bire fringence, 51-63
  isotactic polypropylene,
    57-63, 117, 119-121,
    146, 154
  poly(ethylene terephtha-
    late), 213

Chain folding, 3, 4
Crystal structure, iso-
    tactic polypropy-
    lene, 3-7
Crimp, poly(ethylene
    terephthalate), 224
Crystalline orientation,
    10-15, 20, 24-28,
    37-41, 116-119,
    146-150, 167, 214
Crystallinity, 20-22

Density, 21, 116, 146
Dynamic modulus, isotactic
    polypropylene, 192-
    193
  poly(ethylene terephtha-
    late), 234-237

Elastic recovery, 205-207
Electron diffraction, 82,
    86-89
Electron microscopy, 7,
    12-14, 86-89, 131-
    132, 138, 143

Failure mechanics, 172-198
Fabrication, fibers, iso-

tactic polypropy-
    lene, 141
  poly(ethylene terephtha-
    late), 212-215
film isotactic polypro-
    pylene, 114-140
Folded chains, 3, 4
Fracture envelope, 173-
    177, 195-197

Glass transition tempera-
    ture, 176, 188, 191,
    235-237

Histograms, true stress,
    179-185
Hydroxypropylcellulose
    film, 160, 245

Infrared dichroism, 63-64,
    70-82
Infrared theory, 64-76
Interfibrillar links, 143
Interlammelar anisotropy,
    121-133
Internal field, 59, 62
Intrinsic birefringence,
    57-63
  determination, 57-60
  isotactic polypropylene,
    58
  poly(ethylene terephtha-
    late), 215
Intrinsic Sonic modulas
  determination, 43, 45,
    48-50
  isotactic polypropylene,
    49
  poly(ethylene terephtha-
    late), 238

Lamellae, 3, 7, 10, 114,
    121-133
Light scattering, small-
    angle, 89-109
theory, 89-99

experimental, 89, 90,
    92-94, 99-109
isotactic polypropylene
    Fibers, 151-153
    Films, 100-109, 133-139
Long period, 121-133, 151,
    168-171, 224-228

Microfibrils, 14-15, 132
Micro x-ray diffraction,
    82
Mixed metal systems,
    246-247
Molecular anisotropy, 115-
    121

Nucleation, 7

Orientation function def-
    initions, 20, 28,
    164-165

Polarizability, 53, 54,
    59, 62-63
Polycrystalline polymer,
    15-17, 114
Polypropylene structure,
    3-9, 78-79

Recovery Factor, 205-207
Recovery Tensile, 204-210
Refractive Index, 53-54,
    59
Rod scattering, 245
Rodlike superstructure,
    245

Secondary crystallization,
    9
Segmental motion, 192-193,
    234-237
Shrinkage, isotactic poly-
    propylene, 244-245
poly(ethylene terephtha-
    late, 219-223
Small-angle x-ray scatter-
    ing (SAXS), 121-133

experimental, isotactic
    polypropylene,
    Fibers, 151
    Films, 121-133
poly(ethylene terephtha-
    late), Fibers, Long
    Spacing, 224-228
    intensity, 228-234
Small-angle light-scatter-
    ing (SALS), 89-109
    theory, 87-99
    experimental, 89, 90,
        92-94, 99-109
    isotactic polypropylene,
        Fibers, 151-153
        Films, 100-109, 133-
        139
Sonic modulus, 41-51, 57-
    63, 117, 119, 154,
    237-240, 246-247
Sonic velocity, 44, 46-48
Spherulites, deformed,
    11-14
    fibers, 145-146, 151,
        155
    films, 127, 133-139
    structure, 7-10
Spun fiber, 141-145
Stress-strain behavior,
    161-172

Tenacity, 194-198, 215-
    219
Tensile Modulus, poly-
    (ethylene tere-
    phthalate), 237-240
True strain, 166
True stress, 165-169
Tie molecules, 3, 6, 10,
    114
Transitions, spherulitic-
    Microfibrillar, 14-
    15, 131-132, 140,
    157-158, 164, 172-
    173
    dynamic Mechanical, 192-
    193, 234-237

Transition moment, 68–79
Two-phase model, 3, 17–22

Velocity, sonic, 44, 46–
48

Work recovery, 205–207

X-Ray diffraction, wide-
angle (WAXS) theory,
22–28
orientation determina-
tion theory, 28–32
experimental, 32–37
isotactic polypropy-
lene, Fibers, 37–41,
149
Films, 37–41

poly(ethylene tere-
phthalate), Fibers,
214
small-angle (SAXS), 121–
133
experimental, isotactic
polypropylene, Fi-
bers, 151
Films, 121–133
poly(ethylene tere-
phthalate), Fibers,
long Spacing, 224–
228
intensity, 228–234

Yield behavior, 198–204
Yield point, 162, 171